D1426777
NORTH EAST LINC SHIRE
LIBRARIES
WITHDRAWN
FROM STOCK
AUTHORISED:

Experimental Fluid Mechanics

BY

P. BRADSHAW, B.A.

PERGAMON PRESS

Oxford · New York · Toronto · Sydney · Braunschweig

Pergamon Press Ltd., Headington Hill Hall, Oxford
Pergamon Press Inc., Maxwell House, Fairview Park, Elmsford, New York 10523
Pergamon of Canada Ltd., 207 Queen's Quay West, Toronto 1
Pergamon Press (Aust.) Pty. Ltd., 19a Boundary Street, Rushcutters Bay, N.S.W. 2011, Australia
Vieweg & Sohn GmbH, Burgplatz 1, Braunschweig

First edition 1964
Second edition 1970
Library of Congress Catalog Card No. 75-102402

Printed in Great Britain by A. Wheaton & Co., Exeter

08 006979 7 (flexicover)
08 006981 9 (hard cover)

Contents

Editorial Introduction

THE books in the Thermodynamics and Fluid Mechanics division of the Commonwealth Library have been planned as a series. They cover those subjects in thermodynamics and fluid mechanics that are normally taught to mechanical engineering students in a three-year undergraduate course.

Although there will be some cross-reference to other books in the division, each volume will be self-contained. Lecturers will therefore be able to recommend to their students a volume covering the particular course which they are teaching. A student will be able to purchase a short, low-priced, soft-cover book containing material which is relevant to his immediate needs, rather than a large volume in which most of the contents are outside his current field of study.

This book is intended to supplement other volumes in the series and the titles published to date are given inside the rear cover. It meets the immediate requirements of the mechanical engineering student in his undergraduate course, and should also prove a useful volume for the aeronautical engineering undergraduate, for the graduate student and the research worker.

Preface to Second Edition

THE basic sections of the book have not been greatly changed although I have updated the existing information in many places and inserted details of a few techniques, like the hydrogen bubble method and the laser interferometer, that have come into common use since the first edition was published. A section on automatic data recording and digital processing has been added, but this branch of experimental technique is changing so rapidly that I have deliberately kept it to a sketch.

Some readers of the first edition complained of its aeronautical bias—which I had been at some pains to arrange so as to give coherence—and I have now tried to make the terminology, references and examples more general, not, I hope, at the expense of clarity. I stand by my remarks on pp. 3–4: so much money has been spent on aerospace activities that it would be discreditable if aeronautics had not advanced more rapidly than other branches of the subject.

I have replaced the last section of the book, previously a review of a few published experiments which are now rather dated, by a new chapter on the problems and techniques found in different branches of fluid mechanics: this is intended to give specialists a rough idea of what goes on in the subjects they are *not* studying. I have again included descriptions of some recently published experimental work in a variety of subjects. The reader will find the account of his own speciality superficial and incomplete: he should realize that, of necessity, the other accounts are too.

Acknowledgements

I AM indebted to Mr. N. C. Lambourne for Plate 2 and to Mr. R. J. North for Plate 9.

Figures 7, 16 and 43, and all the Plates except 3, 6, 7 and 8, are published by permission of the National Physical Laboratory.

Figures 6, 47 and 50, and Plate 3, are published by permission of the Controller, H.M. Stationery Office.

Plate 6 is published by courtesy of Dr. J. D. Woods and the Editors of the *Journal of Fluid Mechanics*, Plate 7 by courtesy of the Ballistic Research Laboratories, Aberdeen Proving Ground, Maryland, and Plate 8 by courtesy of Prof. S. J. Kline, Stanford University.

I am grateful to Dr. R. C. Pankhurst and to the Editors for making many helpful comments on the manuscript of this book, and to Sheila Bradshaw for help with the proof-reading.

CHAPTER 1

Introduction

THIS book is intended for students of the theory of fluid mechanics, who must also learn about the physical situations which the theory represents, and especially for those who contemplate specializing in the experimental side of the subject rather than the theoretical side. The experimental techniques which the student will meet in his practical work are fully described, together with a representative selection of the more advanced techniques used in research and development testing. The description of techniques is not intended to be exhaustive, and the major part of the book is made up of an introduction to the phenomena which can be experimentally observed, a discussion of their significance in terms of the equations of motion and of dimensional analysis, and a description of the way in which experiments in fluid mechanics are conducted.

In this introductory chapter, we first discuss the use of dimensional analysis, in particular the way in which it can be used to relate the results of model tests to flows at full scale: next, the simplification of the equations of motion that can be achieved by applying them only in restricted regions of the flow dominated by particular flow phenomena is described, together with examples of how the flow over a body, and the apparatus required for its investigation, depend on the speed of flow relative to the speed of sound. In Chapter 2, wind tunnels are discussed, both because tunnels and other test rigs with similar features are the basic test facilities of laboratory fluid mechanics and because most of the physical and mathematical features of the subject are well illustrated by the flow in wind tunnels. Four chapters on techniques of measurement are followed by a discussion of the conduct of experiments and the writing of reports, and the last chapter is a very rapid survey of specialized branches of the subject. Because few general principles can be laid down to guide the experimenter, he must necessarily learn from particular examples, and the index is intended to be used as a guide to the separate references to each subject made in addition to the main discussion. The page number of each main discussion is printed in bold type in the index: the reader may find it instructive, after reading the main section, to look up the subsidiary illustrative references.

The most important thing for the student to gain is a physical understanding of the behaviour of fluid flow: next in importance is a knowledge of the advantages and limitations of the experimental techniques, and last of all comes a knowledge of the manual skills needed to use the techniques. This being the order of importance, the book is not primarily a work of reference for those actively engaged in experimental work, though it may be useful, particularly to newcomers to the field. It is a supplement to the other volumes in the Thermodynamics and Fluid Mechanics division of this Library and a preliminary to practical experience, but not a substitute for either.

Theory and Experiment in Fluid Mechanics

Compared with some of the other physical sciences, fluid mechanics has a sound and extensive mathematical basis. It is usually possible to formulate the differential equations describing the variation of velocity, pressure, temperature and other quantities of interest, and the problem is completely determined mathematically when the conditions at the boundaries of the flow are specified as well. Unfortunately it has not been possible to *solve* the equations in general, and particular solutions have so far been obtained only for flows in which some of the terms in the equations can be neglected or approximated. Although the flow around aerofoils and other so-called streamlined bodies can be adequately if laboriously calculated by choosing appropriate simplifications or approximations in different regions of the flow, even these "patchwork" calculations frequently require the use of empirical formulae. Accordingly, experimental fluid mechanics is a flourishing science, important both in its own right as a method of solving flow problems with or without theoretical support, and also as a source of information for the improvement of theoretical analyses. It is this interdependence of theory and experiment which makes fluid mechanics a particularly stimulating study. Although this book deals only with the techniques and methods of approach needed for experimental work, the reader should remember that, even when mathematical analysis cannot solve the problem, it may help to indicate which features of the flow are most important. The experimenter's most useful tool is mathematics, and if he neglects to use it he is making just as big a mistake as the theorist who takes no interest in experimental work, although all but the most talented workers find it necessary to concentrate either on experimental or theoretical work.

Aeronautics and Other Branches of the Subject

Experimental fluid mechanics has been developed largely for and by the aircraft industry, and most of the more sophisticated techniques to be described have been pioneered in work on aircraft or gas turbines. At present, workers in other branches of the subject seem to be learning from aerodynamicists rather than vice versa. Non-aeronautical aerodynamics, the study of airflow past earthbound structures, shares the methods if not the problems of aeronautics. We shall see below that the flow of Newtonian liquids is essentially the same as the flow of air and other gases although there are effects peculiar to each, such as cavitation and compressibility respectively: quite frequently experiments on models of ship hulls are done in air, while water tunnels have been used extensively for flow visualization studies in aeronautics. Naval architecture and experimental meterology have many techniques in common with aeronautics. Because of the pre-eminence of aeronautics, however, and because most readers are likely to be interested in this branch of the subject, aeronautical language has been used in this book in places where greater generality would have required awkward paraphrases, and many of the particular examples have been drawn from aeronautics: the student primarily interested in liquid flows should have no special difficulty in appreciating the principles which the examples are intended to illustrate (see Preface to Second Edition).

Relative Motion

Generally speaking it is immaterial whether the fluid moves and the solid boundaries of the flow are stationary or vice versa, so that

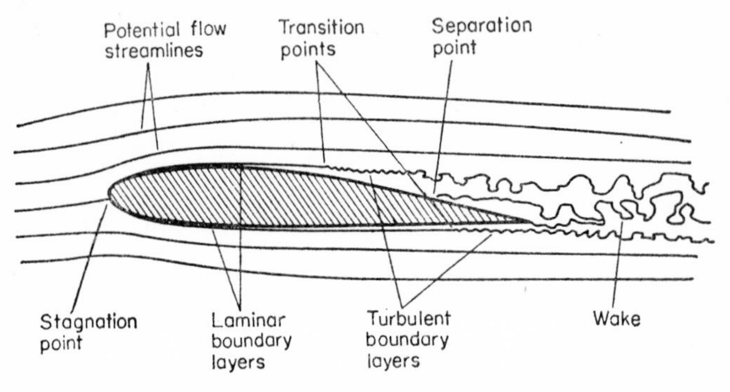

FIG. 1. Flow over a two-dimensional aerofoil.

experiments can be made with whichever arrangement is more convenient. Usually, a stream of fluid is directed past a stationary body, as in a wind tunnel, because this makes it easier to attach force or pressure-measuring instruments to the body, but towing-tank tests and free-flight experiments have their uses. When in what follows we speak of the motion of fluid past a body, without further qualification, the reader may find it helpful to think of a stationary aerofoil in a moving stream as shown in Fig. 1, although most of the phenomena to be described are also present in duct flows and in the passage of towed or self-propelled bodies through stationary fluid.

Historical Development of Fluid Mechanics[1] *

The first extensive scientific investigation of fluid motion was the classical theory of hydrodynamics pioneered by Helmholtz and others. It was assumed that the flow could be described by a potential, like that used in electrical theory, so that the velocity of the fluid at a point corresponded to the gradient of the potential at that point. This assumption is equivalent to the neglect of friction, viscosity and other effects which irreversibly dissipate kinetic energy into thermal internal energy. The outstanding defect of the theory was that it predicted that the drag force exerted by the fluid on a body was zero. This conflict with experiment is a consequence of the assumption of reversibility, because the work needed to propel a non-lifting body relative to a fluid is really dissipated into thermal energy by the viscous shear stress at the surface and in the wake.

The influence of viscosity on fluid motion was first properly realized towards the end of the last century, although Stokes' work on very slow flows entirely dominated by viscosity was done in 1850 at about the same time as the work of Helmholtz. In 1883 Reynolds showed experimentally that the steady motion of a viscous fluid could break down into an irregular turbulent motion of great complexity. At about the same time Mach, in studies of the flight of shells, found that the compressibility of the air greatly affected the motion at speeds comparable with that of sound. Parsons, in his experiments with ships powered by steam turbines, found that local boiling of the water under reduced pressure (cavitation) occurred in the flow round highly loaded propellers. Compressibility and cavitation later became of great importance in aircraft and ship design respectively. In late years departures from perfect gas behaviour in flows at very high temperatures or very low density have become important in the development of missiles, and one of the many examples of the interaction of aeronautics with other subjects is provided by recent studies of the effect of magnetic fields on the flow of ionized (electrically con-

* Superior numbers are those of references (pp. 211–14).

ducting) high-temperature gases, an effect which is important in astrophysics and in research on thermonuclear power generation.

We see that fluid motion is affected by viscosity, compressibility and possibly other phenomena, ignored in the classical theory of hydrodynamics. When these effects are included in the mathematical equations, they become intractable, and experimental work would also be excessively complicated if it were not possible to neglect some of the effects in part or all of a given flow. An analysis of their relative importance in a given case is a necessary preliminary to theoretical or experimental investigation: the chief method of doing this is the method of dimensional analysis.

Dimensional Analysis[2]

Fluid mechanics, unlike, say, atomic physics, does not seek to establish the absolute values of fundamental constants, and most experimental results can be related to flows on a different scale or in a different fluid provided that they are expressed in non-dimensional form by dividing the measured quantity by a reference quantity having the same dimensions of mass, length, time, etc. The technique of discovering the appropriate non-dimensional groups from the known variables is called dimensional analysis.

The easiest way of understanding the basis of dimensional analysis is to note that any equation relating physical quantities must have the same "dimensions" of mass, length, time, temperature, and so on, on each side. If it did not, then a change in the unit of measurement of one of the unbalanced dimensions would produce a change in the equation without any corresponding change in the physical situation which the equation is supposed to represent. Let us consider a well-known apparently unbalanced equation, $U = 66 \cdot 15\sqrt{h}$, which is used to calculate the speed of an airstream from measurements of dynamic pressure. U is the speed in feet per second and h is the dynamic pressure, measured in inches as a liquid level difference on a water manometer. If we measured the level difference in centimetres instead of inches the formula would not apply, not even if we changed the units of velocity to centimetres per second. The numerical factor 66·15 really has hidden dimensions (see eqn. (4)) because it depends on the density of the air, and also on the density of water and the acceleration due to gravity, which determine the liquid level difference produced by an applied pressure: all three of these hidden quantities involve the dimension of length. If the value of any of the quantities were to change (without a change in the unit of measurement) then this change in the physical situation would not be reflected by any change in the equation which is supposed to represent the physical situation.

The use of equations with numerical factors containing hidden dimensions is likely to lead to confusion. If the equation is to represent all the variables in the physical situation it must be expressible as a combination of *non-dimensional* groups: an equation whose terms all have the same dimensions (a dimensionally homogeneous equation) can be converted into non-dimensional form by dividing through by one of the terms. Non-dimensional dependent variables will hereafter be called coefficients and non-dimensional independent variables will be called parameters: this use of the word "coefficient" has no simple affinity with its use in discussing algebraic or differential equations.

Dimensional analysis is an extension of geometrical similarity, in which the only relevant dimension is that of length. Two systems in which *all* the relevant dimensions (including mass, time, temperature, and so on as well as length) are similar are called dynamically similar, or, in the context of dimensional analysis, "similar" without qualification. Complete dynamical similarity must include geometrical similarity.

Consider the flow of fluid past a body, and let the velocity at some reference point (for instance a point far enough from the body for the velocity to be unaffected by the presence of the body) be U_1. If this reference velocity is now increased, to U_2 say, we expect that the velocity at each point in the flow field will increase in the ratio U_2/U_1 provided that the flow is not influenced by any other velocity or by any other quantity having the dimensions of velocity. If the body happened to be a ship driven by a propeller or an aeroplane driven by a jet, the rotational speed of the propeller or the velocity of the jet would also have to be increased in the ratio U_2/U_1 for similarity to be maintained: we shall see below that less obvious quantities with the dimensions of velocity may appear.

Flows past two geometrically similar bodies at the *same* speed will be similar providing that no other quantity with the dimensions of length appears, so that flows at different reference speeds past geometrically similar bodies will also be similar providing that no further quantities with the dimensions of length or of velocity appear.

So far we have only considered the geometrical and dynamical behaviour of the system without taking account of the properties of the fluid. It is not a sufficient condition for two flows to be similar that the fluid properties should be the same for each: in fact it is not even a necessary condition. The necessary and sufficient condition for similarity is that all possible non-dimensional combinations of the geometrical and dynamical scales of the system *and* the properties of the fluid shall be the same in the two cases considered. For instance, we may deduce from the experiments of Mach mentioned in the last section that the speed of sound may be a relevant property of the fluid, so that to ensure similarity the ratio of a reference velocity to the velocity of sound in the fluid must be the same in the two cases. We note that the speed of sound in a perfect gas is $a = \sqrt{(\gamma p/\rho)}$, a quantity which one would not immediately recognize as having the dimensions of velocity: this suggests that a systematic approach will be necessary to identify all the possible non-dimensional groups occurring in a given problem.

Let us write a general, dimensionally homogeneous equation as $f(Q_1, Q_2, Q_3, Q_4, \ldots, Q_n) = 0$. The variables Q each have different dimensions of mass, length and time: a useful notation is to write $[Q] = [M, L, T]$ to indicate that Q has dimensions of mass, length and time; for instance the dimensions of density are given by $[\rho] = [M/L^3]$. We can express the same equation as a functional equation involving terms like $Q_1^{\alpha} . Q_2^{\beta} \ldots Q_n^{\nu}$ where the indices

$\alpha, \beta \ldots \nu$ are not necessarily all non-zero—that is, we admit terms like $Q_1 . Q^3_2$. We can now, by our initial assumption that the equation is dimensionally homogeneous, choose each of the terms so that they all have the same dimensions, or in particular so that they are all non-dimensional. If, for instance, the equation was $Q_1 + Q_2 + Q_3 = 0$ we could rewrite it as $(Q_1/Q_3) + (Q_2/Q_3) + 1 = 0$. The exact arrangement of the combinations of the terms does not concern us: we are only interested in arranging the terms in non-dimensional form. Note that equations containing transcendental functions can still be written, formally, as combinations of products by expanding the transcendental functions in power series. If the equation is to be homogeneous in its dimensions, each of the arguments must be non-dimensional because it will appear raised to several different powers.

We can form non-dimensional groups by equating to zero the dimensions of mass, length and time in the general term $Q_1^\alpha . Q_2^\beta . Q_3^\gamma \ldots Q_n^\nu$, thus obtaining three equations (or more, generally one equation for each of k fundamental dimensions, if we also include temperature, electric charge and so on as dimensions) in n unknowns, the indices $\alpha, \beta \ldots \nu$. *If* all k equations are independent we can eliminate k unknowns in terms of the other n–k indices, leaving n–k unknowns, from which we deduce n–k dimensionless groups.

For example, let us suppose that

$$[Q_1] = [\mathrm{ML/T^2}],\ \ [Q_2] = [\mathrm{M/L^3}],\ \ [Q_3] = [\mathrm{L/T}],\ \ [Q_4] = [\mathrm{L}],$$
$$[Q_5] = [\mathrm{M/LT}].$$

so that

$$[Q_1^\alpha . Q_2^\beta . Q_3^\gamma . Q_4^\delta . Q_5^\epsilon] = [\mathrm{M}^{\alpha+\beta+\epsilon}\mathrm{L}^{\alpha-3\beta+\gamma+\delta-\epsilon}\mathrm{T}^{-2\alpha-\gamma-\epsilon}] = [0]$$

giving

$$\alpha + \beta + \epsilon = 0$$
$$\alpha - 3\beta + \gamma + \delta - \epsilon = 0$$
$$-2\alpha - \gamma - \epsilon = 0$$

If we solve for γ, δ and ϵ in terms of α and β we obtain

$$\gamma = \beta - \alpha$$
$$\delta = \beta - \alpha$$
$$\epsilon = -\alpha - \beta\ .$$

so that

$$[Q_1^{\alpha} . Q_2^{\beta} . Q_3^{\beta-\alpha} . Q_4^{\beta-\alpha} . Q_5^{-\alpha-\beta}] = [0], \text{ and}$$
$$[(Q_1/Q_3\, Q_4\, Q_5)^{\alpha}] = [0] \text{ and } [(Q_2\, Q_3\, Q_4/Q_5)^{\beta}] = [0]$$

We are entitled to say that *each* of the last-named groups has zero dimensions because their product has been shown to have zero dimensions for any values of the arbitrary indices α and β. It is usually convenient to let α and β take the value unity, but it sometimes happens that such a choice for the n–k arbitrary indices leads to non-dimensional groups with fractional powers occurring, in which case it is usual, though in no way necessary, to raise the groups to a high enough power to eliminate fractional indices: it may also happen that the physical significance of a parameter can be more clearly seen by raising it to some particular power. Further, it may sometimes be convenient to multiply a coefficient by some function of a parameter so as to obtain a new definition for the coefficient which minimizes its variation with the parameters, or to combine parameters to form a new (but not independent) parameter.

As an example, consider the force F exerted by a stream of fluid of density ρ and viscosity μ on a sphere of diameter d past which it is flowing with reference velocity U. Then F, ρ, U, d and μ correspond to the variables Q_1, Q_2, Q_3, Q_4 and Q_5 in the above analysis, and our two dimensionless groups are $F/dU\mu$ and $\rho Ud/\mu$. If we regard the force on the sphere as the dependent variable, $F/dU\mu$ becomes a coefficient and $\rho Ud/\mu$ is to be regarded as a parameter on which the coefficient depends, $F/dU\mu = f(\rho Ud/\mu)$. If we regard the force as an independent variable and require a coefficient for the velocity, as we might if we wished to investigate the terminal velocity of the sphere in free fall under a force equal to its weight, we obtain $\rho Ud/\mu$ as a function of the parameter

$\rho F/\mu^2 = (F/\mu Ud) \, . \, (\rho Ud/\mu)$, so that $\rho Ud/\mu = f(\rho F/\mu^2)$. It is general aeronautical practice to base force coefficients on the cross-sectional or plan area of the body and the dynamic pressure of the air stream, and for this purpose the force coefficient $F/\rho U^2 d^2 = (F/\mu Ud)/(\rho/Ud/\mu)$ is appropriate, so that $F/\rho U^2 d^2 = f(\rho Ud/\mu)$.

An alternative way of finding all the dimensionless groups that can be made from a given set of variables is to take one variable, say the dependent variable in which one is most interested, and reduce its dimensions of mass, length and time to zero in turn by multiplying it by other variables. After the process is complete, one repeats it on one of the variables which does not appear in the first dimensionless group, and so on. In the case of the flow past a sphere, we could start with the force F and proceed as follows:

$$[F] = [\mathrm{ML/T^2}]$$

$$[F/\mu] = [\mathrm{L^2/T}] \qquad \text{eliminating mass}$$

$$[F/\mu U] = [\mathrm{L}] \qquad \text{eliminating time}$$

$$[F/\mu Ud] = [0] \qquad \text{eliminating length,}$$

ρ has not been used:

$$[\rho] = [\mathrm{M/L^3}]$$

$$[\rho/\mu] = [\mathrm{T/L^2}]$$

$$[\rho U/\mu] = [1/\mathrm{L}]$$

$$[\rho Ud/\mu] = [0]$$

Hence $F/\mu Ud = f(\rho Ud/\mu)$.

The advantage of this alternative procedure is that one can nearly always identify most of the dimensionless groups from experience or by inspection, which makes the construction and solution of a set of simultaneous equations rather wasteful of time, but the beginner will probably find that the more formal method is easier to use.

Neither method indicates whether the right variables have been used. If a variable has been omitted, this will sometimes show up as an inability to construct dimensionless groups containing all the other variables, but only if the omitted variable has a dimension which occurs in only one of the other variables. If an unimportant variable has been inserted, it may well be possible to combine it into a non-dimensional group expressing the effect it would have if it were important: if, for example, we included the surface tension of the fluid, σ, with the dimensions of force per unit length or $[M/T^2]$, we should obtain the parameter $\sigma/\rho U^2 d$, irrespective of whether a free surface occurred in the problem or not. The point of these statements is that dimensional analysis is merely a mechanical process which only attains its full value if one keeps a close watch on the physical significance of the dimensionless groups.

The physical significance of coefficients is usually obvious, though physical intuition is helpful in deciding on the best reference quantity to use in forming any particular coefficient, for instance in choosing between the three coefficients for F, the force on the sphere, mentioned above. Most parameters in fluid mechanics can be regarded as the ratio of two typical forces, or in particular as the ratio of a typical force to a typical pressure or inertia force. Therefore a parameter represents the relative importance of a typical force and a typical pressure or inertia force. The usual reference pressure in incompressible flow is the dynamic pressure, $\frac{1}{2}\rho U^2$, where U is the reference velocity. In compressible flow, the absolute stagnation pressure is often used: it is related to the other possible reference pressures in terms of the Mach number.

The parameter $\rho U d/\mu$ derived above is called the Reynolds number: the ratio μ/ρ occurs so frequently in parameters that it is given the symbol ν and the name of kinematic viscosity. The Reynolds number can be rewritten $\rho U^2/(\mu U/d)$: recalling the definition of viscous shear stress as the product of viscosity and velocity gradient, we see that the Reynolds number is the ratio of a reference pressure and a typical viscous shear stress and is therefore a measure of the importance, or otherwise, of viscous forces. It will later be shown that the Reynolds number can sometimes be conveniently based not on a geometrical dimension of the body but on the size of the region in which viscous effects are important.

The Mach number U/a is the square root of $\rho U^2/\gamma p$ and is therefore related to the ratio of the dynamic pressure and the absolute pressure, so that it can be regarded as a measure of the importance of compressibility effects. The ratio of specific heats, γ, is a non-dimensional group which frequently occurs in compressible flow formulae; one must not overlook such ready-made non-dimensional groups.

The Mach number and Reynolds number are the most important parameters in fluid mechanics. Other parameters must be introduced when heat transfer, surface tension effects, thermal convection and other phenomena appear, and examples of these parameters are given in the other sections of this book.

It is not necessary to go into the theory of dimensional analysis in any more detail: reference 2 gives a fuller account. It is, however, important for the student to realize the importance of thinking in terms of non-dimensional quantities, of deciding which of the effects represented by the parameters are important and which are negligible in any particular experiment, and of recording results in non-dimensional coefficient form so that they can be easily compared with results obtained on a different scale. Even if it is certain that there will never be any need to make a comparison with results obtained in a different fluid it is still necessary to make the results non-dimensional, or at least to reduce them to standard values of the properties of the fluid, so as to allow for day-to-day variations in temperature, barometric pressure and so on.

Units

Although most experimental results can be expressed non-dimensionally, it is still necessary to pay some attention to the systems of units to be used for recording the measurements of force, pressure and so on: sometimes, the units can be arbitrary if the measured quantity is to be divided by a reference quantity measured in the same way, as in the case of a pressure coefficient equal to the ratio of two readings on a pressure gauge, but where the properties of the fluid are involved, as in the calculation of Reynolds number, the units used must form a self-consistent system.

In the section on dimensional analysis we assumed that the fundamental dimensions of mechanics were mass, length and time, with temperature and certain electrical quantities as additional dimensions. It is certainly the usual practice in pure science to regard mass, length and time as fundamental, but there is no reason, except that of convenience, why we should not define mass, time and velocity as fundamental dimensions and derive a length unit as the product of the time and velocity units: the light-year is such a unit. Alternatively, we could define force as a fundamental dimension instead of mass, and use Newton's second law of motion to derive a mass unit in terms of the force unit and the units of length and time. The reasons for preferring a particular system of dimensions are (i) convenience of definition, (ii) convenience of use.

In English-speaking countries, the foot–pound–second–degree Centigrade system has been used in the past for most aeronautical purposes, the degree Fahrenheit being used in some branches of thermodynamics. The old metric (centimetre–gram–second–degree Centigrade) system has been used in some research work and was in general use on the continent of Europe. Both systems are being superseded by the Système International d'Unités (S.I.) of which the main units are the metre, kilogram, second and degree Celsius or Kelvin). The only unfamiliar unit that frequently occurs in fluid mechanics is the Newton or kg m s^{-2}, the force needed to accelerate a 1 kg mass at 1 m s^{-2}: it is, of course, 10^5 dynes or almost exactly 0·225 lbf. By 1968, the S.I. had been adopted by thirty countries as their only legal system, and the British government is promoting its general adoption in the United Kingdom. Its advantage over the ft–lb–sec system is its freedom from awkward conversion factors, although the properties of matter are bound to have awkward values in any system of units. The disadvantage of the changeover is that in the next few years we shall be subjected to a good deal of scientific pedantry from editors, authors and teachers on such weighty matters as whether one atmosphere is 101·3 kN m^{-2} or $1\cdot013 \times 10^5$ N m^{-2}. Also, students may have difficulty in finding the definitions of obsolete units used in older books and reports: examples are the slug, a unit of mass used in the foot–pound (force)–second system and defined as the mass that would be accelerated at 1 ft sec^{-2} by 1 lbf, and the poise (named after Poiseuille), a unit of viscosity in the centimetre–gram force–sec system equal to 1 gmf $s^{-1}cm^{-2}$, but there are few more in fluid mechanics.

In reprinting this book, most of the dimensions quoted in the text have been altered to S.I. units: references to existing structures such as wind tunnels have been left in the ft–lb–sec system. Conversion of existing dimensions to the metric system should be made with discretion: let us not hear of wind tunnels with working sections 1·2192 m by 0·9144 m when the tolerances on a 4 ft by 3 ft tunnel are at least ±1 mm.

A common practice in fluid mechanics is to measure and quote pressures in units of height of a liquid column, the inch or centimetre of water or mercury being the most common. The purist objection that these units depend on the density of the liquid, which changes with temperature, and on the gravitational acceleration, which depends on geographical position, may usually be ignored because the standard of accuracy expected in experimental fluid mechanics is not so high as in fundamental physics. The following table may be useful for converting from one to another of these odd units of pressure:

one atmosphere equals 10·31 m water
405·8 in. water
760 mm mercury
29·92 in. mercury
$1\cdot013 \times 10^5$ N m^{-2}
14·70 lb force $in.^{-2}$

The most convenient way of reducing results to non-dimensional form when they involve fluid properties is to work out multiplying factors based on fluid properties at standard temperature and pressure, indicated by suffix 0 and usually 15°C and 760 mm Hg for air, and then to allow for changes in temperature and pressure by applying correction factors near unity, obtained from charts, nomograms or tables. As an example, the Reynolds number Uc/ν_0 for air (ν_0=1/69130 m² s⁻¹) is numerically equal to

$$U\,(\text{ms}^{-1}) \times c\,(\text{m}) \times 69130$$

$$\text{or } 12{\cdot}65\sqrt{(\tfrac{1}{2}\rho U^2)}\,(\text{cm water}) \times c\,(\text{m}) \times 69130$$

At other temperatures and pressures, these values must be multiplied by ν_0/ν and $\nu_0/\nu\sqrt{(\rho_0/\rho)}$ respectively, if these factors are sufficiently different from unity to warrant their inclusion. The use of these factors is also necessary when an experiment has to be performed at a constant Reynolds number or Mach number from day to day: this precaution is necessary if the behaviour of the flow changes rapidly with either variable.

The Equations of Motion

The following treatment is intended only as a brief outline for future reference: the equations of motion and the various simplified equations derived from them are described more rigorously in some of the other volumes of this series.

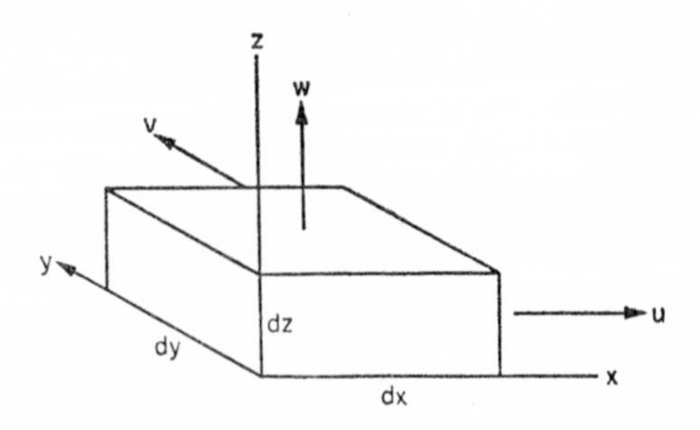

FIG. 2. Notation for equations of motion.

The Navier–Stokes Equations

The Navier–Stokes equations[3] express Newton's second law of motion, the balance of the rate of change of momentum of a given fluid element with the viscous and other forces acting on it. Let u, v and w be the velocity components in the orthogonal directions x, y and z at a given point in the fluid (Fig. 2) and let p be the pressure: in the most general case of compressible, unsteady, three-dimensional viscous flow, the equation expressing Newton's law for the component of momentum in the x direction, is

$$\left.\begin{aligned}
&\text{Inertia force } \frac{\partial u}{\partial t} + u\frac{\partial u}{\partial x} + v\frac{\partial u}{\partial y} + w\frac{\partial u}{\partial z}\\
&\quad\text{equals}\\
&\text{Pressure gradient force } -\frac{1}{\rho}\frac{\partial p}{\partial x}\\
&\quad\text{plus}\\
&\text{Viscous force } \frac{1}{\rho}\frac{\partial}{\partial x}\left(\mu\left(2\frac{\partial u}{\partial x} - \frac{2}{3}\operatorname{div}\mathbf{V}\right)\right)\\
&\qquad + \frac{1}{\rho}\frac{\partial}{\partial y}\left(\mu\left(\frac{\partial u}{\partial y} + \frac{\partial v}{\partial x}\right)\right) + \frac{1}{\rho}\frac{\partial}{\partial z}\left(\mu\left(\frac{\partial u}{\partial z} + \frac{\partial w}{\partial x}\right)\right)\\
&\quad\text{plus}\\
&\text{Body force } F_x,
\end{aligned}\right\} \quad (1)$$

where $\operatorname{div}\mathbf{V} = \frac{\partial u}{\partial x} + \frac{\partial v}{\partial y} + \frac{\partial w}{\partial z}$ and all forces are per unit mass.

The equations for the other two components of momentum are obtained by cyclic interchange of symbols. Usually, x is taken parallel to the surface or to the direction of the stream at large distances from the body. In addition, we have the equation of conservation of mass, or the equation of continuity,

$$\frac{\partial \rho}{\partial t} + \frac{\partial \rho u}{\partial x} + \frac{\partial \rho v}{\partial y} + \frac{\partial \rho w}{\partial z} = 0 \quad (2)$$

Both this and the momentum equation can be derived by considering the flow of fluid into and out of an infinitesimal box-shaped control volume of sides dx, dy and dz, say (Fig. 2). For instance, the force in the x direction exerted on the fluid in the control volume by the pressure gradient is equal to $-(\partial p/\partial x).\mathrm{d}x.(\mathrm{d}y.\mathrm{d}z)$, $\mathrm{d}y.\mathrm{d}z$ being the area of the sides of the box normal to the x direction. Expressed as force per unit mass, or acceleration, this expression becomes $-(1/\rho).(\partial p/\partial x)$.

The boundary conditions, which specify the field in which the equations are to be solved, are (usually) that the velocity at infinity is uniform and that the normal and tangential components of velocity at a solid surface are zero.

The equations are quite intractable unless some simplification can be made. Among the difficulties is the presence of terms like $u\partial u/\partial x$, involving the product of velocities, which means that we cannot add two solutions of the equation together and so construct a third equal to their sum. Also, each equation contains all three velocity components so that the three equations must be considered together.

Incompressible Flow

A great simplification is to assume that the density and viscosity are constant, a good approximation in the absence of large temperature or pressure changes. Then, the continuity equation becomes div $\mathbf{V} = 0$ and the viscous-force term in the x-component momentum equation is

$$\nu \left(\frac{\partial^2 u}{\partial x^2} + \frac{\partial^2 u}{\partial y^2} + \frac{\partial^2 u}{\partial z^2}\right)$$

The inertia and pressure terms are unaltered in appearance. In the special case of two-dimensional flow, $w = 0$ and $\partial/\partial z = 0$.

"Inviscid" Flow and Bernoulli's Equation

Let us now consider "one-dimensional" steady motion without viscous or body forces, putting $\nu = 0$ and taking the x-direction as the direction of motion so that $v = w = 0$. The x-component momentum equation reduces to

$$u \frac{\partial u}{\partial x} + \frac{1}{\rho} \frac{\partial p}{\partial x} = 0 \tag{3}$$

in compressible *or* incompressible flow. If ρ = constant (incompressible flow) the equation can be integrated in the x direction, or more generally in the direction of motion, to give

$$p + \tfrac{1}{2}\rho u^2 = \text{constant} = p_0 \text{ say} \tag{4}$$

In the presence of a body force due to gravity, $F_z = -g$, we have

$$p + \tfrac{1}{2}\rho u^2 + \rho g z = \text{constant.}$$

This is Bernoulli's equation[5], and is valid along any line, which coincides with the direction of motion at each point along its length so that fluid does not cross the line: such a line is called a streamline, and in steady flow it is simply the path followed by a fluid particle (see p. 146). In steady flow the constant p_0 is the maximum pressure which the fluid on the particular streamline considered could attain if brought to rest without the action of any viscous forces, and is called the total pressure or stagnation pressure. p is called the pressure (without qualification) or the static pressure. The difference between the total pressure and the static pressure is called the dynamic pressure and $\frac{1}{2}\rho U^2$, instead of ρU^2, is therefore used as a reference pressure difference in incompressible flow. In practice, the pressure at the front of a body rises at one point, called the stagnation point (Fig. 1) to a value closely equal to p_0 unless the Reynolds number (based, let us say, on the stream velocity and the radius of curvature of the body at the stagnation point) is less than about 100, when the ratio of viscous forces to pressure forces is no longer very small. We see that a measurement of the pressure on the body at the stagnation point, together with a measurement of the static pressure at a point sufficiently far from the body to be unaffected by its presence, can be used to deduce the velocity of the fluid. We must note that although the total pressure is constant along a given streamline it may vary from one streamline to another.

If the density is not constant but the flow is adiabatic and reversible (isentropic) so that we may put $p \propto \rho^{\gamma}$ or $\rho = \rho_1 (p/p_1)^{1/\gamma}$ for a perfect gas, eqn. (3) can again be integrated after substituting for ρ to give

$$\frac{u^2}{2} + \frac{\gamma}{\gamma - 1}\frac{p}{\rho} = \text{constant or } \frac{u^2}{2} + c_p T = \text{constant} = c_p T_0 \text{ say.} \tag{5}$$

Further use of the isentropic expansion law and the expression $a = \sqrt{(\gamma RT)}$ for the velocity of sound leads to[5]

$$\frac{p_0}{p} = \left(1 + \frac{\gamma - 1}{2} M^2\right)^{\gamma/(\gamma-1)} \tag{6}$$

where M is the local Mach number u/a and p_0 is again the pressure reached if the fluid is brought adiabatically and reversibly to rest. If the right-hand side of this equation is expanded in powers of M^2 we find that to order M^2 it is identical with equation (4), because the assumption that M^2 is small is identical with the assumption that density changes are small. $p_0 - p$ is again called the dynamic pressure, although sometimes this title is reserved for $\frac{1}{2}\rho U^2$ which is *not* equal to $p_0 - p$ in compressible flow.

Shock Waves

If the Mach number of the flow exceeds unity, part of the retardation of the stream near the front of a body occurs suddenly across a surface roughly normal to the flow direction, called a normal shock wave (Plate 7). The thickness of the shock wave is determined by the mean free path of the gas molecules and viscous dissipation occurs within it so that there is a loss of total pressure through the shock, but for most purposes the thickness of the wave can be taken as infinitesimal and the change of total pressure and other variables can be calculated from considerations of continuity, momentum and energy, so that measurements of total pressure behind a normal shock wave can still be used for velocity determination.

The Boundary-layer Approximation[3]

Although the simplification of ignoring viscous forces is permissible for the flow up to the stagnation point on a body at a reasonably high Reynolds number, it is not realistic for the flow over the rest of the body. However, if the Reynolds number is high,

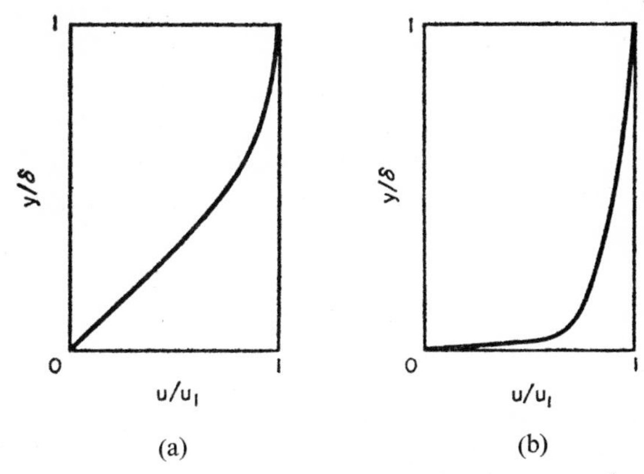

FIG 3. Boundary layer velocity profiles at low Mach numbers with no longitudinal pressure gradient:

(a) Laminar boundary layer.
(b) Turbulent boundary layer.

it is found experimentally that the direct effect of viscosity is confined to a thin layer adjacent to the surface of the body called the boundary layer, in which the velocity rises from zero at the surface to a value very near that predicted by inviscid flow theory at the outer edge (Figs. 1, 3); viscous effects are also important in the wake behind the body which is partly composed of the fluid which has passed through the boundary layers. The external flow is called "inviscid": this term is an abbreviation of "quasi-inviscid" or "unaffected by direct viscous forces". Considerable simplifications of the equations of motion can be made if we assume that the boundary layers and wake are thin and that their streamwise rate of growth is small, so that velocity gradients within the boundary layer in the direction parallel to the surface will be much smaller than velocity gradients in a direction normal to the surface. The only change in the appearance of the x-component momentum equation (1) is that the viscous term reduces further to $\nu \partial^2 u/\partial y^2$ but the equation for the component perpendicular to the surface, the y direction, say, reduces to $\partial p/\partial y = 0$, so that the static pressure within the boundary layer is constant at a given streamwise position and equal to the static pressure just outside the layer. Using Bernoulli's equation, and considering steady, two-dimensional incompressible flow for simplicity, we can eliminate the static pressure entirely by replacing $-(1/\rho)\, \partial p/\partial x$ by $U_1\, \partial U_1/\partial x$ where U_1 is the velocity just outside the boundary layer, assumed to be in the x direction. The x-component momentum equation then appears as

$$u\frac{\partial u}{\partial x} + v\frac{\partial u}{\partial y} = U_1\frac{\mathrm{d}U_1}{\mathrm{d}x} + \nu\frac{\partial^2 u}{\partial y^2} \tag{7}$$

which is a great improvement on eqn. (1), and which can be solved numerically for arbitrary $U_1(x)$.

As an additional benefit, the velocity U_1, usually called the free-stream velocity, can be calculated by extensions of the methods of classical hydrodynamics, because the flow outside the boundary layer and wake can be assumed to be inviscid and of constant total pressure except where it passes through shock waves. The boundary conditions of the outer inviscid or "potential" flow (see p. 5) are not exactly the same as if the boundary layer and wake did not exist, because the fluid in the viscous shear layers is slower moving, and the rate of mass flow is the same as if the velocity were equal to the free-stream velocity down to a distance δ_1 from the surface or from the centre line of the wake and were thereafter zero. The effect on the outer flow is the same as if the thickness of the body were increased by an amount δ_1. The length δ_1 is called the displacement thickness of the boundary layer: it is a function of position on the body.

$$\delta_1 = \int_0^{\delta} \left(1 - \frac{\rho u}{\rho_1 u_1}\right) dy \tag{8}$$

where δ represents the edge of the boundary layer where the integrand falls to zero. We can also define the momentum thickness δ_2 by substituting "rate of momentum flow" for "rate of mass flow" in the above definition, obtaining

$$\delta_2 = \int_0^{\delta} \frac{\rho u}{\rho_1 u_1} \left(1 - \frac{u}{u_1}\right) dy \tag{9}$$

If the value of δ_1 is small, both in the boundary layer and in the wake, the body is said to be streamlined because its surface is very nearly a streamline of the inviscid flow outside the viscous layers: most bodies designed to move through a fluid are streamlined. Bodies whose boundary layers separate from the surface well before they reach the rear of the body, forming a thick wake, are called bluff (Plate 5), and the effect of the viscous layer displacement thickness on the inviscid flow is considerable: sometimes a successive approximation technique can be used to match the solution in the viscous layers and the inviscid flow. Separation occurs only when $\partial p/\partial x$ is positive so that viscous forces and pressure gradient forces both tend to retard the flow near the surface, and finally to reverse it if $\partial p/\partial x$ is large enough.

The line on which the boundary layer leaves the surface is called the separation line: it coincides, very nearly, with the line at which the viscous shear stress at the surface, $\mu(\partial u/\partial y)_w$, becomes zero (Figs. 4, 5) and behind this line the fluid nearest the surface moves slowly forward with respect to the body (as can be seen in Plate 5). The flow near separation and in the wake is usually rather unsteady.

In the neighbourhood of the separation line and in the region of reversed flow the rate of growth of the viscous shear layer is not small and the boundary-layer approximation is not valid, but it can be applied to the wake further downstream. It can also be applied to jets, and to the flow in nearly straight ducts. Although for simplicity we have restricted part of the analysis to steady incompressible flow and said nothing about the motion in the z direction

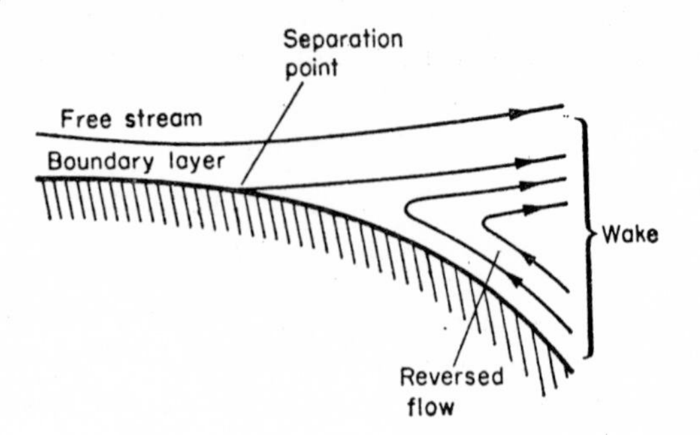

FIG. 4. Boundary layer separation.

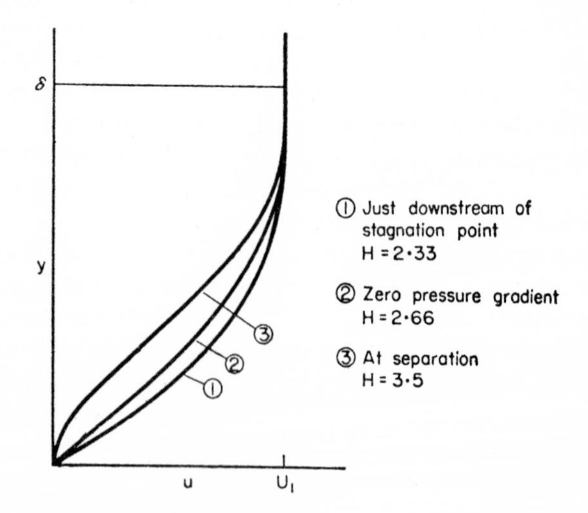

FIG. 5. Laminar boundary layer velocity profiles.

in a three-dimensional flow, the boundary-layer approximation can be applied to any thin shear layer.

The Momentum Integral Equation

A further simplification of eqn. (7) can be made by integrating it with respect to y from 0 to δ. As the velocity gradient at $y = \delta$ is zero, the viscous term $\nu\partial^2 u/\partial y^2$ integrates to $-\nu(\partial u/\partial y)_w$ or τ_w/ρ where τ_w is the shear stress at the surface. Using the continuity equation and substituting δ_1 and δ_2 from eqns. (8) and (9) we obtain

$$\frac{d\delta_2}{dx} = \frac{\tau_w}{\rho_1 U_1^2} - \frac{\delta_2}{U_1}\left(\frac{\delta_1}{\delta_2} + 2 - M_1^2\right)\frac{dU_1}{dx} \tag{10}$$

where we have written the equation in the form which applies to compressible flow: the term M_1^2 vanishes in incompressible flow. $\tau_w/\frac{1}{2}\rho_1 U_1^2$ is the surface shear stress coefficient or surface friction coefficient. Extensions of (10) can be used in three-dimensional boundary layers. We have thus obtained an equation connecting the rate of growth of the boundary layer with the surface shear stress and the longitudinal gradient of free-stream velocity or static pressure. Equation (10) is called the (von Kármán) momentum integral equation: sometimes it is called simply the momentum equation, a title more properly reserved for (7). The ratio δ_1/δ_2 is called the shape parameter or form factor H because it is a measure of the "fullness" of the velocity profile: values of H are marked on the typical profiles of Fig. 5. Equation (10) can be integrated to give δ_2 as a function of x if $\tau_w/\rho_1 U_1^2$ and H can be estimated from empirical formulae.

To give a simple example of this and to give an idea of the magnitude of the boundary layer thickness in a typical case, let us consider the steady laminar boundary layer on a flat plate parallel to the stream, so that dU_1/dx is zero. The shape of the velocity profile, as measured say by the parameter H, is the same at all values of the distance x from the leading edge, so that $(\partial u/\partial y)_w$ is a constant multiple of U_1/δ_2: it follows that $\tau_w/\rho U_1^2$ is equal to a constant times $\nu/U_1\delta_2$, the reciprocal of the Reynolds number based on distance from the leading edge, so that (10) reduces to

$$\frac{d\delta_2}{dx} = k\,\frac{\nu}{U_1\delta_2} \tag{11}$$

which can be integrated immediately. Using a value for the constant k obtained from a solution of the full boundary layer equation (7) (but which might just as well be obtained from empirical data), we find that

$$\delta_2 = 0{\cdot}664\,(\nu x/U_1)^{1/2} \tag{12}$$

If we put $x = 1$ m and $U_1 = 15$ m s^{-1}, so that $U_1x/\nu \simeq 10^6$ for air, we find that δ_2 is about 0·065 cm, and the total thickness of the layer is about 0·5 cm. This latter thickness is difficult to define because the velocity tends asymptotically to the free stream velocity U_1. We note that, in this particular case, the Reynolds number based on boundary layer thickness is a constant times the square root of the Reynolds number based on distance from the leading edge: in circumstances when there is no such simple relation between the two the Reynolds number based on boundary layer thickness is a more useful measure of the importance of viscous effects. A large number of methods is available for calculating the development of a steady, laminar boundary layer beneath a free stream with a prescribed variation of U_1 with x.

Laminar Flow

In the above analysis we have assumed that the flow in the boundary layer is steady. Viscous motion of this sort is called laminar motion, since adjacent laminae of fluid may be imagined to slide over one another. (This is stated here as an explanation for the name rather than as a physical principle.) Unsteady laminar flows can exist, and may be calculated by suitable developments of the methods used for steady flow. In general, the calculation of laminar boundary layers and duct flows has been developed to a satisfactory state, and experimental work on purely laminar flow is not very common, except when the boundaries of the flow are particularly complicated.

Turbulent Flow[6]

Unfortunately, even if the free stream is nominally steady, the flow in the boundary layer at high Reynolds numbers is likely to be eddying and turbulent (Plate 8), particularly if the surface is rough, and the same applies to flow in pipes and channels: wakes and jets become turbulent at very low Reynolds numbers (Plates 5, 6). This turbulence arises because the flow in a laminar shear layer may be unstable, even to infinitesimal disturbances, at the higher Reynolds numbers where the damping effect of viscous forces is small. Inviscid flows are not necessarily unstable in this way, so that the viscosity may help to promote the growth of disturbances and cannot be regarded simply as a source of damping. The theory of laminar flow instability is an interesting extension of boundary layer theory. Experimenters should take to heart the fact that the theory remained unaccepted for about ten years after the appearance of the first papers predicting the main features of the instability, because the wind tunnels of the 1930's had such poor flow that the disturbances were far from infinitesimal, so that the theory did not apply. Later experiments confirmed the applicability of the theory to flows with small disturbances, but our understanding of the later stages of the transition to turbulent flow is still incomplete and, such as it is, is based almost entirely on experiment.

It is not too much to say that nearly all the remaining fluid mechanic problems of flight at moderate speeds, and of the design of ducting, heat exchangers and turbomachinery, are in essence the problems of turbulent motion. Unless special precautions are taken the extent of laminar flow on full size aircraft or ships is small, although large areas of laminar flow may occur on wind tunnel models, the resulting differences in behaviour forming the chief manifestation of "scale effect" as it is called; "Reynolds number effect" is a more informative term. It is usual to provoke transition to turbulence on wind tunnel models artificially to reduce scale effect, by means of trip wires or roughness on the surface near the leading edge.

Although the time–mean velocity in a turbulent boundary layer varies smoothly across the layer (Fig. 3), the velocity at a point fluctuates about the mean value in an almost random fashion as eddies are swept past: the fluctuations may reach ± 30 per cent of the mean velocity in some parts of the layer. If we assume that a turbulent boundary layer or other thin shear layer can still be treated by using the boundary-layer approximation as in a steady laminar layer, and separate the velocity components into a mean (with respect to time) denoted by a capital letter and a fluctuating part denoted by a small letter, the average with respect to time of eqn. (7) for two-dimensional, incompressible flow becomes

$$U\frac{\partial U}{\partial x} + V\frac{\partial U}{\partial y} + \frac{\partial}{\partial x}(\overline{u^2} - \overline{v^2}) = U_1\frac{\mathrm{d}U_1}{\mathrm{d}x} + \nu\frac{\partial^2 U}{\partial y^2} - \frac{\partial \overline{uv}}{\partial y} \tag{13}$$

where we have used the continuity equation $\partial u/\partial x + \partial v/\partial y + \partial w/\partial z = 0$, which is valid for the fluctuations as well as the mean flow, and have also used the y-component momentum equation which is now $p + \overline{\rho v^2} = 0$. Overbars, as usual, denote time averages. Note that the w-component fluctuations do not appear in eqn. (13): even in a two-dimensional mean flow, w-component fluctuations do occur, but do not directly affect the mean motion. The u- and v-component fluctuations, however, exert a great effect on the mean motion. It is found experimentally that, in contrast to the mean velocity components, the three fluctuating components are roughly equal to one another over most of the boundary layer: they fall to zero at the surface and outside the layer.

The Reynolds Stresses

The last terms on each side of eqn. (13) represent extra stress gradients due to the presence of the turbulent flow. The term

$\partial(\overline{u^2} - \overline{v^2})/\partial x$ is usually small compared with the other terms, but the term $\partial\overline{uv}/\partial y$ is far larger than the viscous term $\nu\partial^2 U/\partial y^2$ except very near the surface (see also p. 99). It will be seen that $-\rho\overline{uv}$ has the dimensions of shear stress: it represents the mean transfer in the y-direction of x-component momentum per unit volume ρu. The terms $-\rho\overline{u^2}$ and $-\rho\overline{v^2}$ represent normal stresses. These stresses are known as the Reynolds stresses. The correlation coefficient $\overline{uv}/\sqrt{(\overline{u^2}.\overline{v^2})}$, which would be unity if the u and v fluctuations were identical and zero if they were uncorrelated, can take values as high as $0 \cdot 5$, so that turbulence is by no means as random as it appears and is, unfortunately, capable of producing large shear stresses.

The problem of turbulent shear flow is crystallized by writing the last two terms of eqn. (13) together as $(1/\rho)\ \partial\tau/\partial y$, where τ is the total shear stress, and neglecting the normal stress gradient $(\partial/\partial x)\ (\overline{u^2} - \overline{v^2})$. In this form the equation, and also eqn. (10), applies either to laminar or turbulent flow: to help solve (13) for laminar flow we have the additional equation $\tau = \mu\partial U/\partial y$ relating the shear stress to the mean velocity, but there is no corresponding equation for turbulent flow despite many attempts to derive one on theoretical or experimental grounds. As well as being impossible to calculate, the shear stress in a turbulent boundary layer is usually much larger than in a laminar boundary layer of the same thickness with the same free stream speed, so that the surface shear stress, or "surface friction", τ_w, is larger in turbulent flow, and turbulent boundary layers grow more quickly than laminar ones. As a comparison, the total thickness of a turbulent boundary layer on a flat plate in air at 15 m s^{-1} is about $0 \cdot 023 x^{4/5}$ m, or $2 \cdot 3$ cm at $x = 1$ m where x is measured from the position of transition whereas the corresponding laminar layer is only about $0 \cdot 5$ cm thick, and the surface shear stress or surface friction coefficient $\tau_w/\frac{1}{2}\rho U_1^2$ is about $0 \cdot 0037$ instead of $0 \cdot 00066$. This accounts for the considerable interest in the artificial maintenance of laminar flow by suction or other means.

Turbulent flow, particularly turbulent shear flow, is a largely experimental study. As can be seen from eqn. (13), the mathematical difficulty is that the turbulent fluctuating flow cannot be superimposed on the mean flow without affecting it, which is to say that the equations of motion are non-linear (see p. 18). Because the turbulent motion is almost random it must be studied by taking statistical averages. The statistical mechanics of non-linear, irreversible systems is a subject of quite notorious complication.

As well as controlling the transfer of momentum, turbulence also affects the transfer of heat and matter, and is basic to the study of meteorology and combustion. Most of the experiments on turbulent shear layers have been confined to measurements of the mean flow because the direct measurement of the fluctuations is rather tedious, but even mean flow measurements are complicated by the effect of the turbulence on the measuring instruments. We shall return to this point in later sections.

Duct Flow

At the entrance to a long duct, boundary layers will start to grow in the same way as near the leading edge of an aerofoil. If the entry is well shaped—a so-called "bellmouth"—the boundary layers will remain attached to the sides of the duct and grow until they overlap near the centre of the cross-section. If the entry is not streamlined, the boundary layers will separate and will gradually spread back to the walls further downstream. The flow far downstream will be independent of the entry conditions, except in so far as these may determine whether the flow is laminar or turbulent, and will depend only on the cross-section of the duct if the flow is laminar. If the flow is turbulent the velocity profiles will depend slightly on the Reynolds number, like the profiles in a turbulent boundary layer, because the Reynolds number influences the ratio of the viscous shear stress at the walls to the Reynolds shear stresses nearer the centre of the stream. Theoretical solutions exist for the laminar flow in untapered ducts of various cross-sections, but turbulent flow in ducts of non-circular cross-section is further complicated by secondary flow, the presence of V and W components of *mean* velocity in a nominally parallel flow in the x direction. Secondary flow arises because the wall shear stress and the turbulent intensities vary around the perimeter of non-circular ducts: therefore, stress gradients like $-\rho\partial\overline{v^2}/\partial z$ and $-\rho\partial\overline{w^2}/\partial z$ arise, and must be balanced by accelerations in the z direction. The typical manifestation of secondary flow is the presence of weak longitudinal vortices near the corners of polygonal ducts. If the duct has sharp bends, separations may occur so that any estimate of the pressure drop down the duct depends on experimental data.

Compressible Flow outside Shear Layers

The flow of fluids at high Mach number outside boundary layers and wakes can be described by a single equation for the velocity, obtained by eliminating the density and the pressure from the inviscid momentum equation and the continuity equation, noting that a further relation between pressure and density is obtained from the expression for the velocity of sound, $a^2 = (\partial p/\partial \rho)_s$. The resulting equation for the velocity of the fluid is non-linear, so that solutions cannot be superimposed, which as usual complicates the mathematics considerably. The equation can be linearized by assuming that the velocity only varies slightly from its undisturbed value. Solutions with this restrictive assumption, which also implies that any shock waves occurring must be very weak, are applicable only to slender bodies and wings in completely supersonic flow: although many bodies designed for high speed flight are slender in this sense, the calculations have to be supported by experiments on non-slender bodies. In many cases, particularly the slender delta wing plan form used on many supersonic aircraft, boundary-layer separations may occur by design—or of course by accident—so that inviscid flow calculations must be based on experimental data to describe the boundary conditions. In the transonic region (Mach numbers from about 0·7 to 1·3) where mixed subsonic/supersonic flow occurs over a body (see Plate 9), calculations are very difficult, though progress is being made; the design of aircraft which must penetrate this speed range relies heavily on experiment.

Departures from Perfect-gas Behaviour[7]

At Mach numbers above about five, the temperatures occurring in flight through the atmosphere become so high that "real-gas" effects, including departures from the gas laws, may occur. The temperature may be calculated from a form of the law of conservation of energy. If a gas undergoes an adiabatic process, without doing shaft work but possibly while doing work against viscous forces, the sum of its enthalpy and kinetic energy per unit mass will be constant so that

$$\frac{u^2}{2} + c_p T = \text{constant} = c_p T_0 \tag{5}$$

showing that (5), though *not* (6), applies to an *irreversible* adiabatic process as well as a reversible one: T_0 is the total or stagnation temperature and we have chosen consistent units for c_p. Since $p = \rho T(c_p - c_v)$ and $a^2 = \gamma p/\rho$ for a perfect gas,

$$\frac{T_0}{T} = 1 + \left(\frac{\gamma - 1}{2}\right) M^2 \tag{14}$$

which could also have been deduced from (6) for the particular case of a reversible adiabatic process. At $M = 5$ with $\gamma = 1{\cdot}4$, $T_0/T = 6$ so that if the static temperature is atmospheric, $T_0 =$ 1,500–1,800°K depending on the altitude of flight. The first real-gas effect to occur is usually the variation of specific heats with temperature, so that equations like (14) become inaccurate, although the ratio of specific heats remains more nearly constant than the specific heats themselves. Variations in γ can be caused by molecular relaxation, the failure of molecular vibrations to adjust themselves sufficiently quickly to flow conditions calling for a change in temperature, particularly in the flow through shock waves. Dissociation of molecules will also occur if the temperature is sufficiently high. The experimental and theoretical approaches required for the study of these molecular phenomena when they occur in aerodynamics are being adapted from those used in pure physics, and current progress is rapid: reference to research reports is necessary to obtain an up-to-date picture.

In low-density flows, which are usually associated with high Mach numbers in the case of flight through the upper atmosphere, the ratio of the mean free path of the gas molecules between successive collisions to a typical body or boundary layer dimension may no longer be very small: this non-dimensional parameter is called the Knudsen number. The thickness of shock waves is related to the mean free path and may become large. If the mean free path is not small compared with the length scale characterizing the velocity gradient at the wall, $U_1/(\partial U/\partial y)_w$, the boundary condition that the tangential velocity at the wall is zero (p. 18) must be modified, though the use of the continuum flow equations to describe the rest of the flow may be extended into the low-density range by inserting an empirical or semi-empirical expression for the slip velocity into the boundary conditions. It may be shown from the kinetic theory of gases that the Knudsen number is directly proportional to the ratio of the Mach number to the Reynolds number, and it is now usual to discuss low-density effects in terms of M/R_e to save introducing another non-dimensional number. When the mean free path becomes comparable with the expected boundary-layer thickness, at about $M/\sqrt{R_e} \sim 0 \cdot 1$ when the Reynolds number is based on the length of the body, the boundary-layer concept is no longer useful and the description of the flow depends on experimental results and the kinetic theory of gases. The development of techniques to study low-density flows is in an early stage: the first difficulty is to generate the flow, and some account of the problems involved is given in Chapter 2. In highly rarefied gases where the mean-free path is many times the dimension of the body, molecular interactions can be disregarded and Isaac Newton's assumption that air flow could be treated as the motion of non-interacting inelastic particles comes into its own.

If this account of the equations of motion appears to dwell

tediously upon their limitations it is because we are concerned with the need for experiment in fluid mechanics: the reader should turn to the more rigorous treatment given in the other volumes of this series as a corrective.

The Different Flow Regimes

To sum up the different types of flow which may occur over a body, let us consider the changes which occur in the flow of air over a sphere of, say, 5 cm in diameter at gradually increasing speeds. The numerical value of the Reynolds number Ud/ν is about $300U$ where U is in metres per second.

When the Reynolds number is very much less than unity, so that inertia forces are very much less than viscous forces, the streamlines of the flow are independent of the Reynolds number and the drag force is given by Stokes's formula $D = 3\pi\mu Ud$ so that the drag coefficient $D/\frac{1}{2}\rho U^2(\pi/4)d^2$ is $24/R_e$: the alternative drag coefficient defined on p. 10, equal to the above coefficient multiplied by the Reynolds number, would be a constant. The practical interest in this low Reynolds number range is in the study of the motion of droplets or particles in "aerosols".

At Reynolds numbers of the order of unity Stokes's formula becomes inaccurate because inertia forces start to influence the motion: Oseen in 1910 derived and solved a linearized equation of motion in which inertia terms were approximated but the viscous terms retained in full. This approximation for very low Reynolds numbers is the exact opposite of the boundary layer approximation for high Reynolds numbers, in which the *viscous* terms are approximated.

At $R_e \simeq 20$ the viscous shear layer round the sphere has become thin enough for a boundary-layer separation at the rear to be identified, although the boundary-layer approximation is not at all accurate at such low Reynolds numbers even over the front of the sphere: it always breaks down over the rear owing to the aforementioned separation (see p. 23). The wake behind the sphere becomes unstable at $R_e = 50$ approximately, and apparently vortex rings are shed from the sphere: the corresponding instability in a two-dimensional flow past a cylinder takes the form of alternate shedding of line vortices, similar to those seen in Plate 5, from either side of the body alternately. An extensive study of the flow round a circular cylinder in this Reynolds number range has been made in connection with hot wire anemometry in high speed tunnels. At rather higher Reynolds numbers the flow in the wake becomes turbulent at some distance from the sphere. As the speed is increased, the boundary layer on the sphere becomes thinner and the drag coefficient decreases until it becomes almost constant at about 0·4 at Reynolds numbers of the order of 3000: the drag force is partly composed of surface friction and partly of a net pressure force due to the separation, and resulting incomplete pressure recovery, at the rear of the sphere. The point of transition from oscillatory flow to turbulence in the wake moves nearer to the sphere as the Reynolds number increases, but if the free stream is steady and the surface of the sphere is smooth transition does not occur upstream of the separation point until $R_e \simeq 400{,}000$. However, if the sphere is being tested in a wind tunnel in which the flow is itself noticeably turbulent, transition to turbulence in the boundary layer may occur sooner, at a Reynolds number determined by the intensity of the turbulence and the size of the turbulent eddies compared with the diameter of the sphere. Indeed, the transition Reynolds number, or "critical Reynolds number for spheres", was much used in the past as an index of performance of a wind tunnel. Transition is accompanied by a sharp decrease in the drag coefficient to a value of about 0·1, because a turbulent boundary layer is able to withstand a stronger positive pressure gradient than a laminar layer, so that the region of separated flow at the rear decreases in width, and the pressure on the rear of the sphere rises. It is said that golf balls have dimples so as to disturb the boundary layer and provoke transition: certainly the Reynolds number is in the right range for this to occur. There is also evidence that the flight of a cricket ball may be affected by the seam for the same reason.

If our 5 cm sphere were travelling through water, the kinematic viscosity of which is about 1/13th that of air, it would probably be found that cavitation occurred before the Reynolds number rose into the transition range, so that a foam-filled cavity would form behind the sphere. The onset of cavitation is determined by the relation of the vapour pressure p' to the minimum absolute pressure reached in the flow, so that we expect the non-dimensional ratio $(p_\infty - p')/\frac{1}{2}\rho U^2$, where p_∞ is the pressure far from the sphere, to be an important parameter: it is commonly called the cavitation number.

If the sphere were moving through air, the speed corresponding to the transition Reynolds number of 400,000 would be 120 m s^{-1}, a Mach number of nearly 0·4: this is approximately the Mach number at which shock waves first appear (near the equator) on this very bluff body. No further striking changes occur with further increases in Reynolds number as such so we change to the use of Mach number as the appropriate non-dimensional parameter. The critical Mach number at which shock waves first occur defines the beginning of the transonic range: for shapes of aeronautical interest this speed is at least 0·7 of the speed of sound.

As the Mach number increases through the transonic range the shock wave near the equator strengthens, the drag coefficient rises, and when the Mach number reaches unity, a bow shock wave forms. It is initially very weak and a long way in front of the sphere. On the centre line the shock is normal to the flow by symmetry, but at large distances from the sphere its inclination to the axis becomes $\sin^{-1}(1/M)$, called the Mach angle. The flow continues to change rapidly with Mach number until the rear shock stabilizes coincident with the separation point or, in the case of a streamlined body, at the trailing edge: a normal or nearly normal shock will cause separation of a turbulent boundary layer on account of the static pressure rise through the shock if the upstream Mach number exceeds 1·2 or 1·3 (Plates 7, 9).

At higher Mach numbers, in the fully supersonic range, the flow does not change so rapidly with Mach number. At a Mach number of 5, real-gas effects are likely to be imminent. The power required to propel the sphere at a Mach number of 5 near the ground would be about 150 MW ($C_D \simeq 1$); at high altitudes where the density is low enough to permit economical flight at such high speeds, M/R_e may be high enough for continuum assumptions to be partially or wholly invalid. Very high Mach numbers of the order of 15 to 20 are of course reached only during departure from or re-entry to the Earth's atmosphere: bluff-nosed shapes are commonly used for space vehicles, partly in order to reduce the heat transfer to the sides of the vehicle by causing separation of the boundary layer near the nose.

Categories of Experimental Work

The apparatus and techniques required for experiments tend, like the flow phenomena themselves, to fall into categories depending on the fluid and the speed of the flow, particularly the Mach number. This leads workers to specialize in one of the three main speed ranges, low ("incompressible") speeds or liquid flows, transonic and supersonic speeds, and hypersonic speeds (above $M = 5$ approximately). This is unfortunate, because there is often more connection between flow behaviour at different speeds than between different types of problem at the same speed: for instance, boundary layer behaviour is qualitatively the same at all Mach numbers but requires different experimental techniques from, say, aircraft stability and control. Theoretical workers are more free to pursue an interesting phenomenon throughout the speed range.

This specialization is only a real disadvantage in fundamental research where phenomena, rather than engineering structures, are studied. The other main divisions of research work are project testing, in which the combined effects of various phenomena on idealized versions of engineering structures are investigated, and prototype testing, in which models of the structures themselves are tested and possibly modified on a trial and error basis. We therefore have nine possible divisions of interest, three speed ranges and three types of investigation. The boundaries are, fortunately, not very rigid and contact is fairly close between universities, Government establishments and industrial firms, which tend to specialize in the three types of approach in the orders mentioned. The student should take notice of these boundaries only when they are forced upon him: although at the present rate of progress it is impossible for any one person to keep up to date with all branches of fluid mechanics and still have time left to do original work, one frequently finds that advances in another branch of the subject can be applied to one's own particular interest.

Because the subject is overshadowed by the aircraft industry it is tempting for any account of it to be written in aeronautical terms and to pay particular attention to the supersonic and hypersonic speed ranges which are currently the most important in aeronautical research: however, practically all non-aeronautical fluid mechanics is concerned exclusively with flow at fairly low Mach numbers, and most of the remarks about phenomena and techniques in the following chapters will be made, for simplicity, in the context of incompressible flow. Extensions to compressible flow will be considered separately unless they are quite obvious. The experimental techniques used in non-aeronautical work and not shared with aeronautics are usually highly specialized ones. Some examples are given in Chapter 8 but the list is not intended to be complete. Sometimes great ingenuity is displayed in the application of new techniques to special problems. For instance, one of the most interesting results of turbulence theory concerns the distribution of energy among the various eddies in the range of wavelengths very much smaller than the geometrical scale of the flow but still large enough for the dissipation of eddy energy by viscosity to be negligible: if this range is to be observable experimentally the geometrical scale must be much larger than that obtainable in the laboratory. A recent experiment on this subject was carried out by towing a measuring instrument (a hot-film anemometer) on a minesweeping paravane behind a ship in a tidal channel on the west coast of Canada.

The student who wishes to obtain more information about the experimental techniques outlined in this book should consult reference 8, parts of which are brought up to date by reference 9. A comprehensive treatment of the theory underlying the measurement techniques is given in reference 10, which lays emphasis on the thermodynamic measurements not mentioned here, and a companion volume[11] describes high-speed wind tunnel design and methods of measurement.

Examples

1. A stone is dropped into a pond: write down the non-dimensional parameters on which the subsequent motion of stone, air and water may depend, and state which are likely to be negligible.

2. Re-derive Mach number and Reynolds number as relevant non-dimensional parameters in the flow over an aerofoil of chord c (see p. 4), by writing down the general product $U^{\alpha}c^{\beta}a^{\gamma}\rho^{\delta}a^{\epsilon}$, equating its dimensions of mass, length and time to zero to obtain three simultaneous equations in five unknowns, and eliminating three of the unknowns to obtain the product of Reynolds number and Mach number each raised to an arbitrary power.

3. Derive the continuity equation in its steady incompressible form $(\partial u/\partial x) + (\partial v/\partial y) + (\partial w/\partial z) = 0$ by equating the flow rates into and out of a box-shaped control volume of sides dx, dy, dz.

4. Why does the boundary-layer approximation break down at a separation point but become valid again further downstream in the wake?

5. Given that the Knudsen number λ/l is proportional to M/R_e, and that the viscosity μ is independent of pressure and roughly proportional to $T^{0.76}$ for air, deduce the dependence of the mean free path λ on temperature and pressure.

CHAPTER 2

Tunnels and Test Rigs

MORE power is needed to drive a stream of fluid past a stationary object than to propel that object through the air, or over the water, but it is nearly always much more convenient to make measurements on a stationary object, irrespective of whether the object is designed to move or not. Whirling arms (the term is self-descriptive) and railways have occasionally been used for measurements in air: most investigations of the wave-making resistance of ships are still made in towing tanks, where the required speeds are only of the order of 10m s^{-1} or so, but nearly all laboratory experiments are done in wind tunnels, water tunnels or other test rigs in which the fluids circulate.

A wind tunnel is a device for blowing a steady, uniform (or designedly otherwise) stream of air over a model or full-size structure placed in the working section: the other components of the tunnel serve to generate this uniform stream. Many water tunnels have also been built, chiefly on a small scale for flow visualization purposes, although the Pennsylvania State University has a water tunnel with a 48-in. diameter working section and the Admiralty Research Laboratory a 30-in. water tunnel: the design of water tunnels[12] and wind tunnels is the same in nearly all essentials, and the occasional use of other fluids in tunnels and test rigs has presented no additional difficulties except in the structural arrangements.

One of the simplest wind tunnels is at the Royal Aircraft Establishment: it consists of a box with a number of commercial ventilating fans at one end and a nozzle at the other, and is used for simulating the effect of a wind gust on models which are carried past the end of the nozzle on a railway. The uniformity of the flow is surprisingly good. At the other end of the spectrum of R.A.E. wind tunnels is the 8 ft supersonic tunnel, with a maximum speed of 2·7 times that of sound in a working section 8 ft square, and absorbing 60 MW. Even larger tunnels are in use for tests at transonic speeds and for tests of vertical take-off aircraft, because in both cases the disturbance field of the model extends for large distances normal to the flow direction. A supersonic tunnel large enough to take a full-size aircraft would be prohibitively costly, but several tunnels have been designed or adapted for tests of working gas turbines and ramjets in a supersonic stream. The Arnold Engineering Development Center (U.S.A.) Propulsion Wind Tunnel has a 16 ft square working section and main drive motors totalling 160 MW: a further 58 MW is used for the auxiliary suction plant for transonic operation (see pp. 45–47).

Most of the wind tunnels used in teaching establishments and by industrial firms are of the low speed type, restricted to speeds well below that of sound, and their few high-speed tunnels are less impressive from the engineering point of view. However, most of the features of fluid flow can be demonstrated perfectly well with very simple equipment, and with care the results of tests on a small scale can be applied quite accurately to full-scale flight.

Other types of laboratory test rig have the same general features as wind tunnels, though the elaborate devices for producing uniform flow described below are not always necessary.

Wind Tunnel Layouts

Closed-circuit Tunnels

Figure 6 shows the layout of a typical low-speed (100 m s^{-1}) wind tunnel[13] of the closed-circuit type[14], in which the same air is recirculated. The stream is turned, usually in four 90 deg steps, by rows or "cascades" of closely spaced vanes. There is usually a small vent somewhere in the circuit so that the static pressure in the tunnel does not drift as the air heats up during the run. In low-speed tunnels this vent or "breather" is usually at the rear of the working section, so that the pressure in the settling chamber, where the cross-sectional area is usually at least ten times the cross-sectional area of the working section, is higher than atmospheric by nearly the full dynamic pressure. Sometimes the settling chamber is vented to atmosphere instead to relieve the structural load on this large section: the working section static pressure is then below atmospheric, with the disadvantage that air rushes into

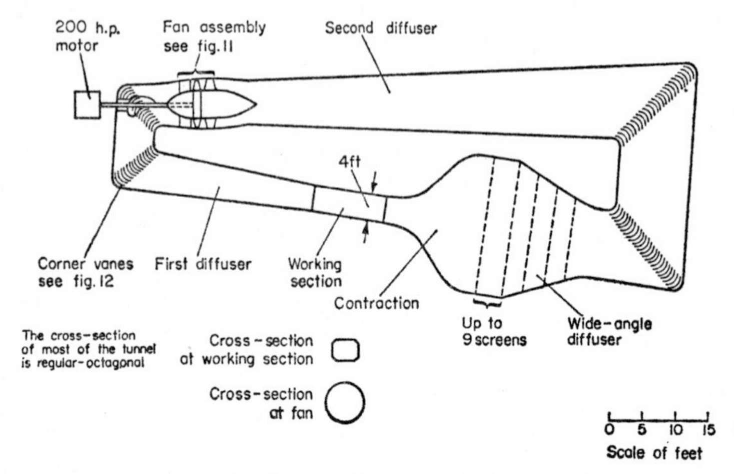

FIG. 6. R.A.E. 4 × 3 ft low-speed wind tunnel.

the tunnel through any holes made for the insertion of model support struts and other impedimenta, unless special care is taken. The R.A.E. 4 × 3 ft tunnel can be run with a vent either in the working section or between the third and fourth corners. In the latter case the working section is enclosed by a sealed observation chamber.

Open-circuit Tunnels

If instead of having only a small vent to atmosphere the tunnel discharges its whole flow to atmosphere at the end of the diffuser and draws in fresh air at the entrance to the settling chamber, it is called an open-circuit tunnel (Fig. 7: the perforated diaphragm of this tunnel improves steadiness at very low speeds). As the kinetic energy of the air being discharged is usually negligible compared with the kinetic energy of the air in the working section, the power required by an open-circuit tunnel may be less than that required by a closed-circuit tunnel of the same aerodynamic design because no power is wasted in the drag of the corner vanes. However, open-circuit tunnels which take in air from the atmosphere are sensitive to draughts, and as a rule only low-speed tunnels, whose compactness and cheapness is more important than their performance, are of this type. Some high-speed intermittent-running tunnels consist

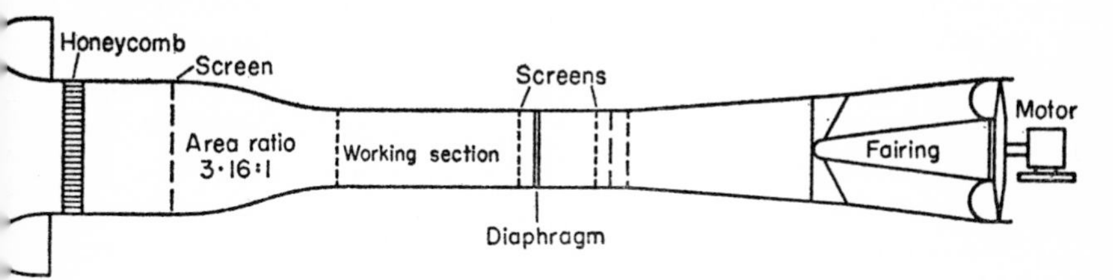

FIG. 7. N.P.L. 18 in. low-speed tunnel.

of a passage between two reservoirs at different pressures: frequently one of the reservoirs is the atmosphere.

Pressure Tunnels

Completely sealed closed-circuit tunnels, which can be pressurized or partly evacuated, can be used to economize on power. The power required to drive a tunnel is proportional to the rate of flow of kinetic energy in the working section, which is equal to $\frac{1}{2}\rho U^2$, the kinetic energy per unit volume, times the rate of volume flow UA, where A is the cross-sectional area. The Reynolds number of a model in the tunnel is proportional to ρU times a typical linear dimension, say $\sqrt{A}$. If we consider two tunnels of the same size, one using atmospheric air and the other air at n times atmospheric pressure, we see that the power required by the latter to produce a given Reynolds number is $1/n^2$ times the power required by the former. Alternatively, a high-speed tunnel can be run at a given Mach number at reduced power by reducing the pressure: the Reynolds number is reduced also, but is usually of secondary interest to Mach number. "Low density" tunnels are designed to run at low pressures to simulate conditions at high altitude.

Power Requirements and Circuit Arrangements

If the air leaving the working section of a tunnel were discharged straight to atmosphere without any attempt to recover its kinetic energy in the form of pressure energy, the power needed to drive the tunnel would be equal to, or slightly larger than, the rate of flow of kinetic energy $\frac{1}{2}\rho U^3 A$. The diffuser is intended to reduce the velocity of the air with the least possible loss of total pressure. The effect of losses here and elsewhere in the tunnel is measured as the ratio of the power needed to drive the tunnel to the rate of flow of kinetic energy through the working section, called the power factor. Power factors as low as 0·2 have been achieved in some low-speed tunnels: the power factor of the R.A.E. 4×3 ft tunnel is about 0·3. The diffuser also serves to reduce the speed of the flow to a convenient value before entry to the fan or compressor which drives the tunnel and then (in a closed-circuit tunnel) to reduce it further before entry to the settling chamber, which generally contains a grid with long tubular streamwise cells, called a honeycomb, and one or more wire gauze screens to reduce the turbulence and spatial irregularities of the flow passing through them. These smoothing devices would absorb a great deal of power if they were placed in a high-speed air-stream. The settling chamber is followed by the contraction, in which the flow is accelerated again to the working-section speed, further reducing disturbances.

It would reduce the power required to an absolute minimum if the expansion to maximum area were rapid, but there is a limit—unfortunately an ill-defined limit—to the pressure gradient which the boundary layer on the diffuser walls will stand without separating, and since the static pressure remains almost constant downstream of a separation there would be a loss of energy as well as considerable unsteadiness of the flow. The overall dimensions of the tunnel circuit are dictated chiefly by the design of the diffuser, so that tunnel designers feel impelled to use as rapid an expansion as appears to be safe: the position of the first corner vanes and thus the overall length of the structure is decided by the need to make the total length of the return leg, comprising the fan or compressor section and the second diffuser which follows it, equal to the length of the leg containing the settling chamber, contraction, working section and first diffuser. We shall see later that the standard of flow in the working section of a tunnel generally improves in proportion to the amount of care taken in designing a gentle diffuser and providing suitable smoothing devices—in other words, tunnels with a cramped return circuit are likely to have poor flow characteristics. As an example, the 13 × 9 ft low-speed tunnel at the National Physical Laboratory, which was designed before the war, is 130 ft in overall length, whereas the R.A.E. 13 × 9 ft tunnel, commissioned in 1954, is about three times this length and has very much better flow.

Transonic and Supersonic Tunnels[15]

There is no fundamental difference between low-speed tunnels and tunnels designed to produce a transonic or supersonic stream, since the flow speed is in the low subsonic range in all parts of a high-speed tunnel except the vicinity of the working section. Supersonic tunnels have a convergent–divergent nozzle ahead of the working section (Fig. 8), in which the flow is accelerated to sonic speed at the throat (provided that the power input to the tunnel is large enough) and reaches the required supersonic speed at the end of the diverging portion (actually, within the rhombus shown in Fig. 8). The Mach number of the flow depends only on the ratio of the cross-sectional area at the nozzle exit to that at the throat, in the manner shown in Fig. 9. The ratio of the static pressure in the working section to the total pressure is therefore also determined by the nozzle area ratio, and increasing the power input to the tunnel does not affect the flow in the working section at all, unless the total pressure alters, which it will not do if the tunnel is vented to atmosphere at the settling chamber or has its pressure controlled by other means. The power required to accelerate the tunnel to supersonic speed is greater than that required for continuous running at low supersonic Mach numbers, because when the flow through the throat is just subsonic it decelerates again downstream of the throat, the boundary layers separate from the diverging walls and large total-pressure losses occur.

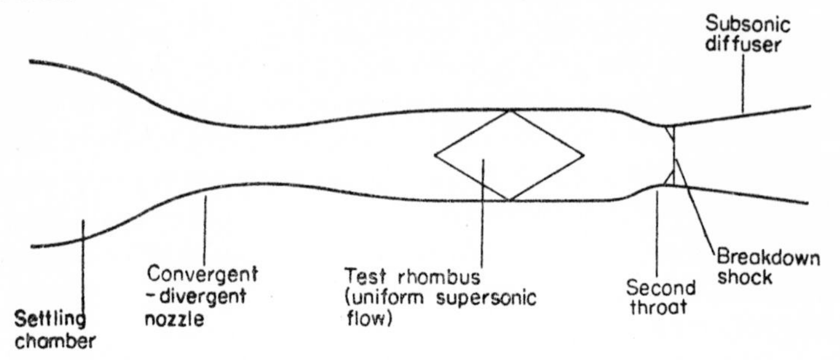

FIG. 8. Supersonic tunnel working section ($M = 2$ nozzle).

At Mach numbers near unity the required nozzle area ratio is very nearly one, the rate of change of Mach number with respect to area becomes very large, and the use of solid-wall convergent–divergent nozzles become impractical: even if a uniform flow could be obtained, the presence of a model in the working section would form a second throat in which the Mach number would automatically return to unity, with gross disturbances to the flow round the model. This phenomenon is known as choking, and for many years prevented tunnel tests being made at all between Mach numbers of about 0·95 and 1·05. The problem has been solved[16] by fitting transonic tunnels with a converging nozzle, similar to the contraction of a low-speed tunnel, leading directly into a working section with slotted or perforated walls surrounded by a plenum chamber (Fig. 10). Since the Mach number is a unique function of the pressure ratio p_0/p (eqn. 6), any desired Mach number can in principle be obtained by sucking air out of the plenum chamber until the static pressure in the plenum chamber is equal to the desired static pressure in the working section.

First considering the case where the desired Mach number slightly exceeds unity we see that the pressure at entry to the working section, where the Mach number will be unity, will be above the plenum chamber pressure so that air will flow out of the working section, thus producing the effect of a diverging nozzle, until at a point further downstream the pressure will be equal to the pressure in the plenum chamber and no further outflow will occur. A uniform flow at the required Mach number will continue to the end of the working section. If a model is now installed, any shock wave generated by it will be reflected from the solid part of the wall as another shock wave and from the apertures in the wall, which form a constant pressure boundary, as an expansion wave. The open-area ratio of the wall can be chosen so that the two types of reflection cancel out. Secondly, if the required Mach number is just below unity the generation of a uniform stream is straightforward, and the insertion of a model merely produces a compensating flow through the walls into the plenum chamber: an almost constant pressure is maintained on the wall, so that the free-stream Mach number remains almost constant as required.

Diffusers

Subsonic tunnels have monotonically diverging diffusers as seen in Figs. 6 and 7, but supersonic tunnels, in which a divergence would produce a further increase in Mach number, are equipped with a second throat at the end of the working section (Fig. 8). Here the flow is decelerated to sonic speed or slightly above; downstream of the throat the Mach number rises again until a shock wave or waves produce a reduction to subsonic speed. It may be shown that a shock wave in the converging portion of the second throat would be unstable and in practice the breakdown shock

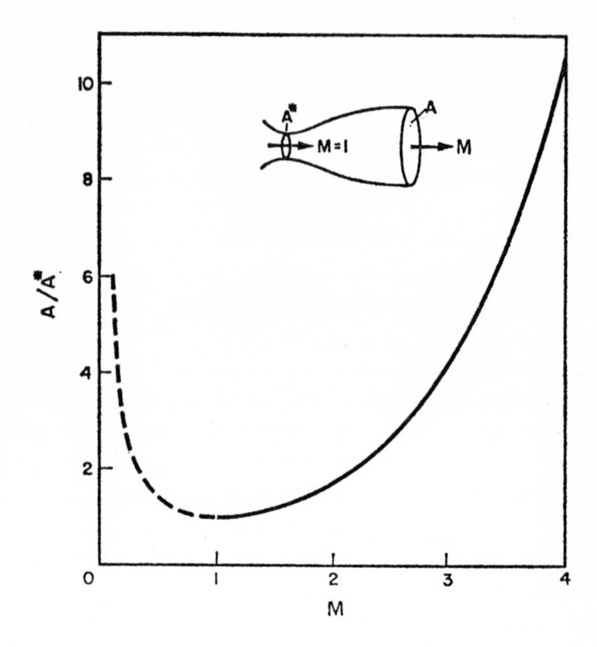

FIG. 9. Ratio of exit area to sonic throat area for supersonic nozzles.

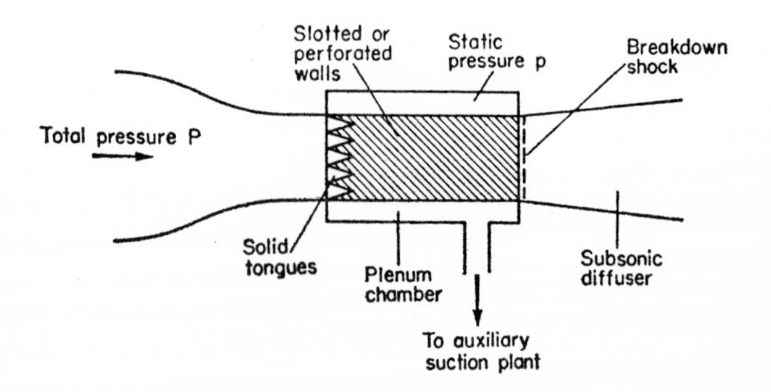

FIG. 10. Transonic tunnel working section.

system is located sufficiently far downstream of the throat to ensure stability under all operating conditions. It seems to be impossible to achieve a deceleration of a uniform stream from above the speed of sound to below it without the occurrence of shock waves.

In tunnels for moderate supersonic speeds a second throat may be omitted, and the flow allowed to break down through a normal shock wave at the full working section Mach number, but at Mach numbers greater than about two the total-pressure loss through a normal shock becomes unacceptably large. The second throat can conveniently be formed by the incidence-changing quadrant used to support sting-mounted models (see p. 127): it is found that the second throat must be rather larger than the first throat in order that supersonic flow can be established in the working section as the tunnel is started up, so that to achieve the best possible diffusion in tunnels for high supersonic speeds the second throat is closed in after the tunnel is started, by adjusting the shape of the tunnel walls.

Subsonic Diffusers

The usual rule of thumb for subsonic diffuser design is to use a conical expansion of 5 deg included angle, but the ideal shape is probably a gradually decreasing rate of expansion: the rate of expansion should in any case depend on the thickness of the (turbulent) boundary layer at the beginning of the diffuser, because we expect the behaviour of boundary layers in adverse pressure gradient to depend on the non-dimensional ratio of a typical pressure gradient force per unit length to a typical surface shear force per unit length, say $\delta(\mathrm{d}p/\mathrm{d}x)/\tau_w$, where δ is some representative boundary layer thickness (see eqn. 10). In the absence of any theoretical information on the ideal shape, conical diffusers are used for structural simplicity. The 5 deg angle is found to be small enough to avoid separation providing that the entry boundary layer is not too thick or the diffuser too long.

It is unusual to have an area ratio of more than three at the fan: a well-designed fan restores a more uniform velocity profile, and further expansion can begin downstream of it. If a further expansion ratio of more than three or four is required, the final stage usually takes place at a rapid expansion with screens (Fig. 6). Although the flow may separate from the walls of the rapid expansion the extent of separation is limited by the screens, which smooth out velocity variations at the expense of a static pressure drop from one side of the screen to the other, as will be explained below. There may be a static pressure drop through the rapid expansion but as the local velocity is so low the extra power expenditure is not important and the reduction in total length of the tunnel is a structural advantage.

Fans and Compressors

Most tunnels are driven by axial-flow fans which produce a static pressure rise (with no appreciable change in axial velocity or dynamic pressure unless the pressure rise is comparable with the absolute pressure) at one point in the tunnel to compensate for the total-pressure losses in the rest of the circuit. The design of tunnel fans covers a wide range[17]: lightly loaded fans, which usually have a high ratio of tip speed to axial velocity and a correspondingly high relative velocity at the blades, produce the required pressure rise with a fairly small blade area and look very much like aircraft propellers. However, fans with high tip speeds cause a great deal of vibration if the approaching stream is not uniform over the cross-section, and a tip speed of more than 150 or 200 m s^{-1} in air implies a relative Mach number approaching that at which shock waves occur, again resulting in noise and vibration.

In modern tunnels, therefore, the fan tip speed is kept as low as possible, not more than two or three times the local axial velocity, and the blade arrangements more closely resemble the axial flow compressors used in gas turbines, with a stator row in front of the rotor (Fig. 11). Most high-speed tunnels have multi-stage compressors because the power factor tends to be higher and the pressure rise may be several times the static pressure at entry. Because it is necessary to return to uniform, non-swirling flow in a circular or polygonal section downstream of the fan, the diameter of the central nacelle is usually a smaller fraction of the fan diameter than in multi-stage compressor practice, and rarely exceeds 0·5 to 0·6 of the fan diameter. As a result the distance, s, between adjacent blades, measured around the circumference, varies considerably from root to tip. The space/chord ratio s/c (see Fig. 11) typifies the fan loading, and it also determines the allowances to be made for the effect of one fan blade upon the adjacent ones.

Fan Design Methods

The usual design method for axial flow compressors, which have fairly small values of s/c, lying between about 0·5 and 1·5, is the camber line method. The difference between the circumferential velocities of the flow at entry to, and exit from, the rotor at each radius r is calculated from momentum considerations and a knowledge of the pressure rise required across the stage. Usually the circumferential velocity at exit from the rotor of a single-stage compressor is chosen to be zero, the swirl induced by the torque on the rotor being cancelled by the stator blades (or pre-rotation vanes as they are often called in wind tunnel practice) mounted upstream. Since the axial velocity of the fluid, U, and the rotational speed of the rotor, $2\pi rn$, are known (Fig. 11) the magnitude and direction of the velocities V_1 and V_2 relative to the rotor blades at entry and exit can be found. Empirical rules are then used to design the blade shape, usually based on a circular-arc camber line, to produce the required deflection of the flow ("cascade design").

These empirical rules are only valid in the range of space/chord ratios from 0·5 to 1·5 for which they were derived: they are not, therefore, very suitable for the design of single stage fans for low-speed tunnels, where the space/chord ratio is rarely less than unity at the root and therefore rises to at least two at the tip if, as is usual, the chord is constant. The camber line method is customarily used for the design of heavily loaded fans for high-speed tunnels, whether or not more than one stage is required.

Low-speed tunnel fans are therefore usually designed by calculating the lift coefficient required at each radius (which is very simply related to the flow inclination angles used in the camber line method), choosing an aerofoil section, and then finding the required angle of incidence (with respect to the mean flow direction relative to the blades) from tests on the same aerofoil section in an infinite stream. This calculation method is similar to that used for the design of aircraft propellers and, in fact, all fans of

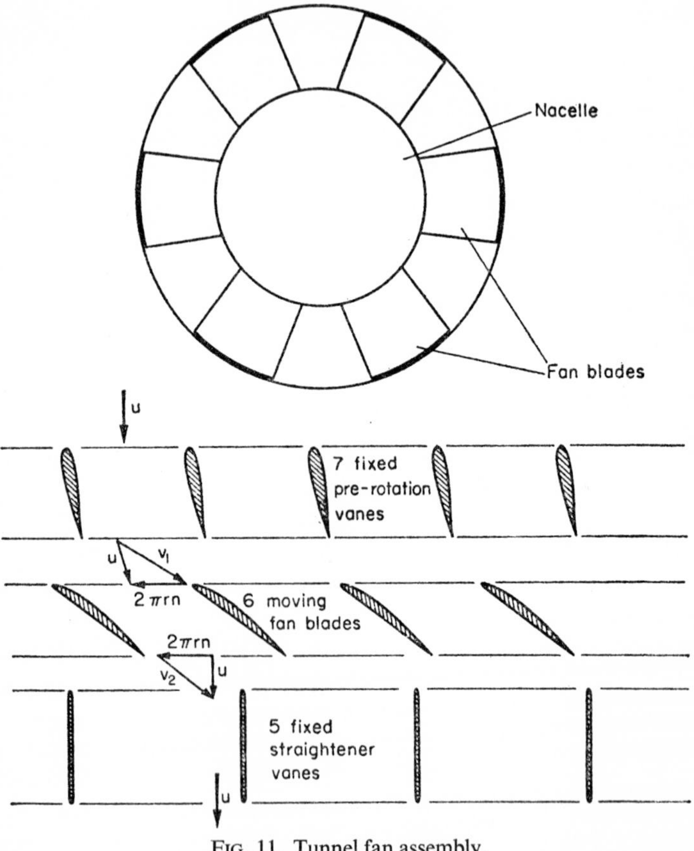

FIG. 11. Tunnel fan assembly.

high space/chord ratio. Corrections to the isolated-aerofoil data are required if the space/chord ratio is less than about two at any radius, but the corrections can be made with sufficient accuracy for space/chord greater than unity. "Centrifugal" blowers and pumps are frequently used for driving small tunnels and test rigs for gas and liquid flow. Modern designs can be at least as quiet and nearly as efficient as axial-flow machines and they maintain their quietness and efficiency over a wide range of flow conditions, but the outlet flow is often rather non-uniform.

Drive Arrangements

Fans are usually driven by electric motors, different supply arrangements being used for different sizes. In subsonic tunnels the drive motor is usually a direct-current shunt-wound type for ease of speed control. Rectifiers can only be used for fairly small d.c. motors, below 50–100 kW, though the size of commercial solid-state controllable rectifiers is continually increasing. Larger motors are usually supplied from a constant-speed d.c. generator whose output voltage is varied by changing the field excitation voltage: the d.c. generator is driven by an a.c. motor supplied directly from the mains. This system, the Ward–Leonard drive, provides such good speed control and economy that the mechanical complication is accepted for the larger sizes of motor. Induction or synchronous motors can be used to drive the fan directly if the tunnel speed is only to be altered over a small range, and are more common for supersonic tunnels where continuous control of motor speed is not required. Occasionally, a.c. motors are used with variable-pitch fans. Electric power is the most convenient, even for the largest sizes, but internal combustion engines and water turbines have also been used.

Small test rigs of a cheap or temporary nature have been driven by commercial blowers, ventilating fans, vacuum cleaners and even aircraft superchargers. Many laboratories have a large-capacity compressed-air supply intended for driving high-speed intermittent tunnels as described in the next section, and which can conveniently be used for continuous running of low-speed tunnels and test rigs. The water-flow equivalent is a tank in the roof.

Intermittent-running Tunnels

The power requirements of continuous-running high-speed tunnels are considerable, and the cost per unit of electrical power may be rather high because the percentage of time per day for which "continuous-running" tunnels are actually being continuously run tends to be small compared with most industrial loads. Accordingly, many establishments have intermittent wind tunnels driven by compressed air from storage bottles which are continuously charged by low-power compressors (blow-down tunnels) or which discharge from atmosphere into an evacuated vessel (suck-down tunnels). The latter type has the advantage that the static pressure in the working section does not vary during the run, providing that the flow is supersonic somewhere between the atmosphere and the vacuum vessel: in this case the final reduction to reservoir pressure is accomplished by a supersonic expansion and a consequent reduction of total pressure in multiple shock waves and turbulence in the vacuum vessel. Blow-down tunnels require an automatic control valve.

An alternative method of using a high-pressure air supply is to drive the tunnel with an injector pump: in this device the required input of momentum is provided by a jet of air blown downstream, either through a single orifice on the centre line of the tunnel or, better, through an annular slot round the wall (Fig. 43). The usual position for the injector is at the beginning of the subsonic diffuser so that the annular jet cleans up the boundary layers at the end of the working section and improves the efficiency of the diffuser: a centre-line jet would have the opposite effect. The working section static pressure of a supersonic "induction" tunnel of this type remains constant throughout the run providing that the tunnel is supplied from atmosphere or vented to atmosphere in the return circuit so that the total pressure remains constant: it is usual to adjust the injector slot width during the run, as the storage pressure decreases, so that the momentum injected is not wastefully larger than that required to run the tunnel. The efficiency of an induction tunnel is usually greater than that of a blow-down tunnel, in which a great deal of energy is dissipated in the throttle valve in the early part of the run when the storage pressure is much higher than the required tunnel total pressure: however, the Mach number of an induction tunnel is limited to about 1·8 by a tendency to choking at the injector position owing to the rapid expansion of the inducing flow.

Corners

Closed-circuit tunnels need corners with guide vanes to deflect the flow without the boundary-layer separations that normally occur at a sharp bend. The space/chord ratio is usually about 0·25 for a 90 deg corner: these values imply that camber-line cascade design methods are not applicable, but the current practice is to use circular-arc camber lines of 85–86 deg subtended angle, with a leading edge angle of attack of 4 or 5 deg and trailing edges set along the centre line of the tunnel (Fig. 12), which have

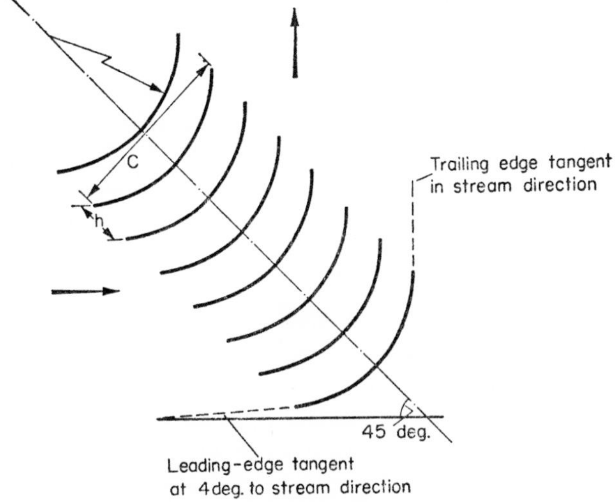

FIG. 12. Circular-arc corner vanes.

been tested once for all by Salter[18] on a small scale. Vanes can be pressed from sheet metal with sharpened edges. The total pressure drop through a corner of this sort is only about 0·08 of the dynamic pressure, compared with a drop equal to or greater than the dynamic pressure in the case of a square elbow: thin vanes have proved just as successful as the thick aerofoil-section vanes used formerly. Attempts have been made to combine corner vanes with an expansion in area round the corner: the efficiency of diffusion is low and there is a danger of flow separation. The only other departure from the conventional 90 deg corner is the use of 180 deg "racecourse" bends with multiple sets of guide vanes, largely for structural reasons, in some American pressure tunnels.

Contractions and Nozzles

Many contractions, including the unusually large contraction of the R.A.E. 4 $\times$ 3 ft wind tunnel, have been designed by eye with or without the precaution of model tests, but it may be deduced from a study of the perfectly satisfactory and very short two-dimensional contraction shown in Plate 4 that one's eye is not necessarily an accurate guide to optimum design. Several mathematical methods, none of them ideal, are available for the design of two-dimensional or axisymmetric contractions in incompressible flow. The flow can be assumed to be essentially inviscid, since in the absence of separations the boundary layers will be very thin and if separations do occur the contraction will obviously be unacceptable, so that we have to solve Laplace's equation

$$\frac{\partial^2 \phi}{\partial x^2} + \frac{\partial^2 \phi}{\partial y^2} + \frac{\partial^2 \phi}{\partial z^2} = 0 \qquad (15)$$

where ϕ is the velocity potential, whose gradient at a point is equal to the velocity at that point: as mentioned on p. 5, the existence of a velocity potential implies that the flow is inviscid. Since $U = \partial\phi/\partial x$, $V = \partial\phi/\partial y$ and $W = \partial\phi/\partial z$, eqn. (15) is a restatement of the incompressible form of the continuity equation (2),

$$\frac{\partial u}{\partial x} + \frac{\partial v}{\partial y} + \frac{\partial w}{\partial z} = 0 \tag{16}$$

Rather than to calculate the velocity distribution produced by a plausible-looking wall shape, it is usual to choose the velocity distribution on the centre line of the contraction, or possibly on the wall, and to calculate the resulting wall shape, using the conditions that the flow must be uniform far upstream and downstream. It can be shown that if the velocity on the wall increases monotonically along the length, then the contraction will have infinitely long and gradual entry and exit contours: if the contraction is to be of finite length, negative velocity gradients must occur at each end, with a resulting risk of boundary layer separation. This is the basic difficulty of contraction design, and even if it were possible to specify the velocity on the contraction wall as a direct function of distance along the wall there would still be uncertainties about the boundary layer behaviour.

A number of methods involve the specification of the velocity on the centre line of the contraction in terms of distance along the centre line: this function is always chosen to be monotonic or neighbouring streamlines would bulge, but all the non-axial streamlines, any one of which can be chosen as the wall, will have regions of negative streamwise velocity gradient which increase in severity with increasing distance from the centre line. Since the length has been specified, the length/diameter ratio decreases as the "wall" streamline is chosen further and further from the centre line, and a compromise between undue length and excessive adverse velocity gradient must be made. Cohen and Ritchie[19] describe a fairly general method of this sort for the design of axisymmetric contractions.

The two-dimensional problem is much easier, and many of the techniques for solving Laplace's equation developed in other branches of engineering have been applied to contraction design. As well as methods in which the velocity on the centre line or the wall is specified, there are now several computer programs for calculating the potential field given the wall shape.

Since the electrical potential in a uniform conductor also obeys Laplace's equation, analogues of fluid flow can be set up. For instance, if the side view of a two-dimensional contraction is cut out of conducting paper with electrodes along the curved walls, the equipotential lines will coincide with the streamlines and the velocity distribution along the walls can be deduced.

Boundary-layer Behaviour

Once the contraction has been designed, it is necessary to estimate whether or not the boundary layer is likely to separate: unless there is special need to make the contraction as short as possible it is quite easy to avoid separation. A difficulty arises if the actual contraction is neither axisymmetric nor two-dimensional but, as is usual in wind tunnels, rectangular in cross-section, with or without fillets in the corner and with or without a change of aspect ratio of the rectangle between the wide end and the narrow end. There is no analytical method for adapting an axisymmetric design to a non-circular cross-section: the usual procedure is to make the area of the rectangle equal to the area of the circle at each station, and to provide fillets in the corners to avoid excessive boundary-layer growth and to mitigate the effects of cross-flow from one wall to another (there is no guarantee that the inviscid-flow streamlines will follow the corners and virtually no hope that the boundary-layer streamlines will do so).

Effect of a Contraction on Velocity Variations

The effectiveness of a contraction in reducing variations of axial velocity over the cross-section can be seen by applying Bernoulli's equation (4) to an incompressible flow with a small region of increased velocity as shown in Fig. 13. The total pressure of the main stream is $p_1 + \frac{1}{2}\rho U_1^2$ and the total pressure in the high-velocity region, which is assumed to have the same static pressure p_1, is $p_1 + \frac{1}{2}\rho U_1^2 . (1 + \delta)^2$. At the exit, the main stream velocity is approximately nU_1 (we neglect the extra mass flow in the high-velocity region) and the static pressure is therefore $p_1 + \frac{1}{2}\rho U_1^2 (1 - n^2)$. The velocity in the high-velocity region is now determined from this value of static pressure and the known value of total pressure, which is not altered by the essentially inviscid flow through the contraction: the velocity is $U_1\sqrt{(n^2 + 2\delta + \delta^2)}$ or approximately $nU_1(1 + \delta/n^2)$. As the velocity at entry is $U_1(1 + \delta)$, we see that the contraction has reduced the percentage velocity variation by a factor n^2. The effect of a contraction on unsteady

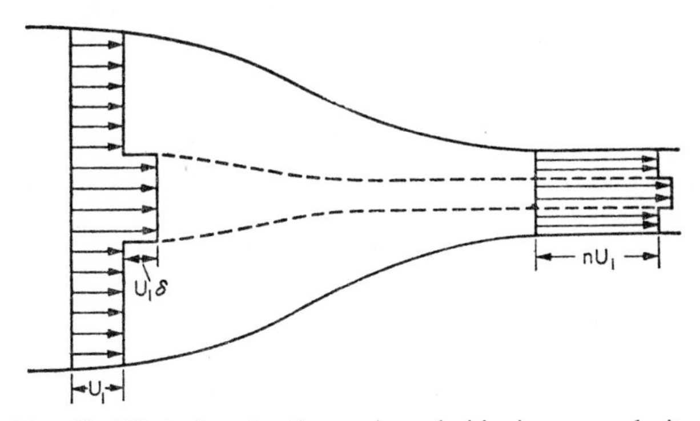

FIG. 13. Effect of contraction on irregularities in mean velocity distribution.

velocity variations and turbulence is more complicated: the reduction of u-component (axial) fluctuations is greater than that of v-component (transverse) fluctuations.

Nozzles for High-speed Tunnels

The convergent part of a convergent–divergent nozzle is a contraction, and it usually follows a permanent contraction which remains the same for all Mach numbers. The difficulty in design is that allowance has to be made for the effects of compressibility as the flow passes through the converging nozzle: the Mach number will rise from a low value at entry to unity at the throat.

These allowances are usually made by assuming that the Mach number is uniform over any given cross-section of the nozzle, that is, that the flow is "one-dimensional". Although this assumption would be rather crude for the initial design of a nozzle—it would fail to predict the necessary presence of regions of adverse velocity gradient on the walls of a monotonically converging nozzle—it has been found to be adequate for making compressibility corrections to an existing design.

In view of the approximate nature of the compressibility corrections, the initial design of the nozzle for incompressible flow is usually carried out approximately or empirically, sometimes even on the assumption of one-dimensional flow, which has been found to be acceptable, when reinforced by experience, providing that the wall angle of the contraction does not exceed about 40 deg.

The compressibility corrections for one-dimensional flow are obtained as follows:

from the continuity equation $\rho UA =$ constant or

$$\frac{\mathrm{d}\rho}{\rho} + \frac{\mathrm{d}U}{U} + \frac{\mathrm{d}A}{A} = 0 \tag{17}$$

we obtain

$$\frac{\mathrm{d}U}{U} = \frac{\mathrm{d}A}{A}(M^2 - 1) \tag{18}$$

where we have used eqn. (3) in the form $U\mathrm{d}U + \mathrm{d}p/\rho = 0$, and the expression $a^2 = (\partial p/\partial \rho)_s$ for the velocity of sound (p. 31), to give $\mathrm{d}U/U + (a^2/U^2)(\mathrm{d}\rho/\rho) = 0$, and eliminated $(\mathrm{d}\rho/\rho)$ between this equation and eqn. (17).

This equation can now be used to determine the actual velocity distribution in compressible flow through the converging nozzle. The Mach number M can be estimated with sufficient accuracy as U/U_{throat} where U_{throat} is equal to the speed of sound at the throat. If the velocity distribution in incompressible flow is satisfactory and no boundary-layer separations occur it is probable that the behaviour in compressible flow will also be acceptable.

Very near the throat, where the Mach number is nearly unity, the flow is dependent only on the local radius of streamwise curvature of the wall, so that the Mach number distribution over the cross-section at any station near the throat is a function only of the ratio of the radius of curvature to the nozzle height, and a number of methods exist for calculating it. The shape of the curve on which the Mach number becomes unity, the sonic line, is required as a starting point for the calculation of the shape of the diverging portion of the nozzle of a supersonic tunnel.

Design of Supersonic Divergent Nozzles

Nearly all supersonic tunnels have two-dimensional nozzles, partly for constructional reasons: not only is the nozzle itself easier to make, but it is also preferable that the windows in the working section shall be flat if any of the methods of observing density changes, described in Chapter 6, are to be used. Another reason is that the only method of calculating axisymmetric nozzle shapes is an adaptation of the method of characteristics used for two-dimensional flow and described below. This method is an approximate numerical one, and the amount of calculation needed to obtain sufficiently accurate results for axisymmetric flow is great[20]. Any errors which do occur, resulting in a non-uniform Mach number distribution over the exit plane of the axisymmetric divergent nozzle, will produce larger non-uniformities on the axis of the nominally uniform and parallel flow downstream of the exit. This is because the disturbances in Mach number existing in the exit plane propagate downstream on "characteristic" lines at the Mach angle $\sin^{-1}(1/M)$ and eventually disturbances in Mach number from all round the circumference are focused on the axis. This focusing effect does not occur in two-dimensional flow, so that the calculation does not have to be carried out to so great an accuracy and presents fewer computational difficulties.

The method of characteristics for two-dimensional flow[21] may be most easily explained as a step-by-step procedure, based on the fact that a small change in the inclination of a wall past which a fluid is flowing at supersonic speed is propagated into the stream at the Mach angle (Fig. 14). In order to accelerate a flow from sonic speed, or slightly above it, to a Mach number of, say 2, a large number of small changes in wall angle, each corresponding to a step in the calculation, must be used. A two-step nozzle, with wall angles greatly exaggerated, is shown in Fig. 14(b). The final Mach

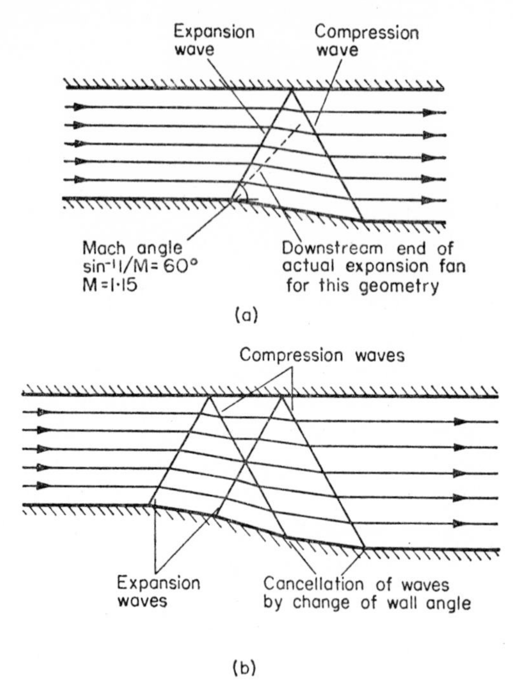

FIG. 14. Mach-wave characteristic lines in supersonic flow.

number depends on the maximum wall angle and also on the number of times the Mach waves are allowed to reflect from the curved wall before being cancelled. The calculation of a nozzle shape still requires a judicious choice of the shape of the convex part of the wall and it is usual to choose a shape which is continuous up to at least its third derivative (though it is approximated in the step-by-step calculation as a series of straight lines). A sudden *large* change in wall angle would produce a disturbance of finite strength, an expansion fan in the case of a sudden divergence, which would not propagate as a single disturbance at the Mach angle and so could not be cancelled by a single change in wall angle further downstream.

Screens

As we saw above, a contraction in area of a stream reduces disturbances by increasing the velocity without altering the total pressure. Since a screen or other grid in a constant-area passage experiences a drag force and therefore reduces the total pressure of the flow passing through it without altering the average velocity, we expect velocity variations to be reduced in this case as well, because the drag force will be greater in regions where the velocity is higher than average, thus tending to equalize the total pressure over the cross-section. The analysis is more complicated than for a contraction, and we will give it only in outline.

If we regard the screen as an idealized resistance, which produces a static pressure drop proportional to the local dynamic pressure, and again consider a region of slightly increased velocity in a flow with uniform static pressure far upstream of the screen, we see that the static pressure just downstream of the screen is lower in the high-velocity region than in the rest of the cross-section. Far downstream of the screen, assuming that the passage remains of constant area, we expect the flow to redistribute itself over the cross-section until the static pressure becomes uniform, so that the velocity at each point will depend on the total pressure, which we assume to remain constant on any one streamline downstream of the screen. We cannot immediately calculate the total pressure in the high-velocity region, because we have no right to assume that the velocity remains constant all the way up to the screen: in fact we should expect the static-pressure disturbance caused by the impingement of the high-velocity region on the screen to have some upstream influence. If we equate the total force on the screen to the total rate of change of momentum of the fluid, and equate the change of total pressure on a streamline in the high-velocity region to the drop in static pressure through the screen, we find that the excess velocity (in the high-velocity region) at the screen is the mean of the excess velocities far upstream and far downstream, and that the ratio of the excess velocity far downstream to the excess velocity far upstream is $(2 - K)/(2 + K)$ where K is the screen pressure-drop coefficient, $\Delta p/\frac{1}{2}\rho U^2$. A more refined analysis by Batchelor[22], which treats a variation in mean velocity as a special case of turbulent flow, with infinitely long eddies, and which takes account of the transverse velocities produced by the static-pressure disturbances, predicts that the excess velocity will be eliminated for $K \simeq 2 \cdot 8$.

Pressure-drop Coefficients of Screens

In practice the pressure-drop coefficient varies with airspeed (that is, with Reynolds number) so that either of the conditions for elimination of the disturbance would only be satisfied at one tunnel speed, and it is usually found that the arrangement of screens is decided by the need to reduce the turbulence generated in the return circuit. To give the maximum reduction of turbulence, the aggregate pressure-drop coefficients of all the screens should be as high as possible: very dense screens are sensitive to imperfections of weave, which cause variations of pressure-drop coefficient from point to point and can produce large variations in mean velocity, so that screens of pressure-drop coefficient greater than about three are avoided. Accordingly, several screens in series are needed to reduce turbulence to an acceptable level, and mean-velocity variations are therefore almost eliminated: those that do remain are largely the result of weaving imperfections or wrinkles in the last screen. Of course, if the test rig is not required to have a particularly low turbulence level, the screen arrangement may be dictated by the requirements of uniformity of mean velocity.

The pressure-drop coefficient K depends on the Reynolds number and the open-area ratio of the screen. Test results for different wire gauze screens of square weave, with open-area ratios typical of wind tunnel practice, can be correlated by plotting an equivalent drag coefficient based on the wire diameter and the average velocity through the pores of the screen against a Reynolds number based on the same variables. Since the drag coefficient defined in this way depends only on the wire Reynolds number and not explicitly on the open-area ratio, it follows that the effect of one wire on the flow past another is also independent of the open-area ratio in the range covered by the tests. Since this equivalent drag coefficient is *not* the same as the drag coefficient of an isolated wire, the independence does not cover the whole range of open-area ratios up to unity (the value for an isolated wire). This is an unusually clear special case of a general theorem—do not extrapolate empirical data! To give some idea of numerical values, a square mesh gauze with 20, 0·0148 in. dia. (28 s.w.g.) wires per inch has a pressure-drop coefficient of about two at air speeds of the order of 7 m s^{-1} at atmospheric pressure.

Honeycombs

Screens, like contractions, reduce the longitudinal components of turbulence or mean-velocity variation to a greater extent than the lateral components, so that the number of screens to be used is determined by the acceptable *lateral* component disturbance in the working section. The lateral components of mean velocity and of the larger turbulent eddies can be reduced more effectively by a honeycomb: the mode of action of a honeycomb with cells elongated in the flow direction is qualitatively obvious but few tests have actually been made, and all that is certain is that the cell length of the honeycomb should be at least six or eight times the cell diameter. A honeycomb naturally produces some turbulence of its own, with eddy sizes of the same order as the cell diameter, which decays very much more slowly than the small-scale turbulence produced by screens. The early wind tunnels which had a honeycomb but no screens (and usually a very small contraction ratio also) suffered from a very high turbulence intensity in the working section although the mean flow was often quite good.

Acceptable Values of Stream Turbulence

As was mentioned in the discussion of boundary layer transition, high turbulence causes confusion in the interpretation of experimental results, and low-frequency unsteadiness is always objectionable. Tunnel designers can never be sure that the tunnel will not be needed for tests in which transition phenomena are important, and so some care is usually taken to reduce turbulence and unsteadiness to an acceptable level in modern tunnels: a root-mean-square u-component fluctuation of 0·1 per cent of the mean velocity is often quoted as an adequately low value for most experiments. The best low-turbulence tunnels have a u-component r.m.s. intensity of the order of 0·02 per cent at low speeds: the intensity of all three components rises with increasing speed, partly because of vibration and partly because the viscous decay of the small-scale turbulence shed by the screens requires a longer distance at the higher speeds where the Reynolds number based on eddy size and r.m.s. intensity is higher. Some idea of the effectiveness of screens and a large contraction in reducing turbulence is given by the performance of the R.A.E. 4×3 ft tunnel (Fig. 6): the turbulence level in the working section is about 0·02 per cent at 30 m s^{-1}, whereas the turbulence upstream of the screens is about 5 per cent.

The turbulence-reducing arrangements in supersonic tunnels are similar but usually less comprehensive. It has now become clear that the high fluctuation intensities measured in supersonic tunnel working sections, rising from 0·4 per cent at $M = 3$ to 1 per cent at $M = 5$ in a typical case (the Jet Propulsion Laboratory 18×20 in. tunnel[23]), are caused by sound waves or weak shock waves radiated from the wall boundary layers, which completely obscure the effects of turbulence in the flow coming from the settling chamber.

Condensation in High-speed Tunnels

If the stagnation temperature in a tunnel is near atmospheric, say 300°K, the static temperature in the working section will be $300/(1 + 0\cdot2M^2)$°K if $\gamma = 1\cdot4$. Even in tunnels of only moderate Mach number this temperature is lower than the dew point of atmospheric air, and careful drying is needed to prevent condensation. At Mach numbers of 3 or 4 the air is close to liquefaction at 90°K, and heating it only permits small increases in Mach number before constructional difficulties become apparent, since the tunnel walls rise to approximately the stagnation temperature. Other diatomic gases (that is, those with $\gamma = 1\cdot4$) have roughly the same boiling point as oxygen and nitrogen, and it is necessary to use helium ($\gamma = 1\cdot66$) to achieve the Mach numbers of 15 to 20 needed to simulate re-entry conditions: it is usually possible to allow approximately for the effects on the flow of a change in γ. Although helium tunnels have been successfully operated, the low static temperature is still a disadvantage because the static temperature experienced by a body moving through the upper atmosphere is of course the atmospheric temperature which, although low, is still high enough for the stagnation temperature to be in the range where the flow is greatly affected by departures from perfect-gas behaviour, both in point of thermodynamic behaviour and by the onset of dissociation and ionization.

Shock Tubes[24]

Several intermittent-running devices which generate a high-speed, high-temperature flow for a very short time go by the general name of shock tubes. The simple shock tube consists of two chambers, one at high pressure and the other more or less evacuated, separated by a diaphragm (Fig. 15). Upon rupture of the diaphragm, a shock wave progresses into the low-pressure chamber where the model is situated, and is followed by a region of steady, high-temperature flow, at a speed which depends on the pressure ratio across the diaphragm but which is supersonic in all cases of interest. The steady flow is terminated by the arrival of the contact surface between the high-pressure and the low-pressure gas, behind which the flow is turbulent. The progress of events is best understood by reference to the "x, t diagram" shown in Fig. 15, in which are plotted the loci in space-time of the various flow discontinuities. The time for which the flow at the model position is steady and suitable for testing increases with the distance between the diaphragm and the model, but so do the non-

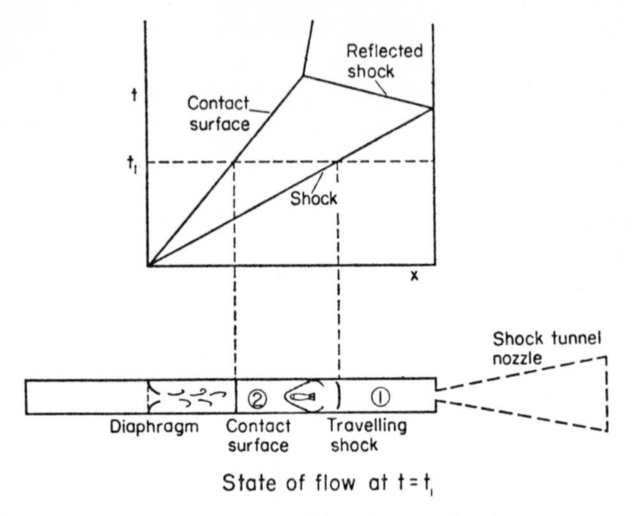

FIG. 15. Flow conditions in a shock tube.

uniformities of the shock wave induced by the effects of viscosity at the walls, and it is unusual for the distance to exceed 100 diameters. The flow is usually defined by the shock Mach number (based on the speed of sound in the undisturbed gas ahead of the shock wave), which depends on the pressure ratio and on the properties of the driver and driven gas if these are different. The Mach number of the steady flow induced by the shock is much less than the shock Mach number. For instance, a shock with a Mach number M_s of 10 in a perfect gas with $\gamma = 1{\cdot}4$ produces a flow

Mach number M_2 of $1 \cdot 83$ at a stagnation temperature 34 times the initial gas temperature—roughly 10,000°K. If we consider a shock tube using hydrogen as the driver gas and air as the driven gas we find that a pressure ratio of 9,600 is required to produce this shock Mach number. The maximum flow Mach number, corresponding to infinite shock Mach number and pressure ratio, would be $1 \cdot 89$ according to perfect-gas theory for $\gamma = 1 \cdot 4$. The driver-gas pressure is limited by strength considerations: pressures up to 1000 atm have been used. The driven-gas pressure is limited by vacuum pumping techniques and the inevitable out-gassing of solid surfaces which continues for some time after evacuation, and a practical limit is set by the need to maintain a reasonably high static pressure in the uniform flow after the passage of the shock so that the Reynolds number shall not be too low: the static pressure in the $M_s = 10$ flow would be 116 times the initial pressure of the driven gas.

Before going on to discuss methods of increasing the flow Mach number we will briefly consider the time for which the flow over the model is steady, that is, the time between the arrival of the shock wave and the arrival of the contact surface. This time is clearly equal to $l/U_2 - l/U_s$ where l is the distance between the diaphragm and the model, U_s is the shock speed and U_2 is the contact surface speed, which is nominally equal to the speed of the uniform flow behind the shock. In the numerical example above, $U_s = a_1 M_s = 3300$ m s^{-1} and $U_2 = a_2 M_2$ which can be calculated, on the assumption of inviscid flow, as 2800 m s^{-1}. If $l = 10$ m, the testing time is $0 \cdot 6 \times 10^{-3}$ sec so that conventional recording methods are not suitable for measurements in shock tubes. It so happens that methods for increasing the flow Mach number can also be adapted to give somewhat increased running times, but a great deal of work has been required to develop suitable measurement techniques.

Shock Tunnels

A much higher flow Mach number can be obtained at the cost of a decrease in static pressure by expanding the (supersonic) flow through a divergent nozzle (Fig. 15), and a shock tube so equipped is called a shock tunnel. In contrast to wind tunnel practice (see p. 61) the nozzle is usually axisymmetric. In practice not all the primary flow is passed into the nozzle. The diameter of the parallel part of the tube must be at least 1/100 of the length of the tube, and the length must be fairly long to give a reasonable testing time: our 10 m tube would have a diameter of at least 10 cm, and expansion of the whole flow through an area ratio of anything up to 1000 (giving a flow Mach number of about 13·5 with an initial Mach number of 1·9) would result in a working section diameter of 3 m, inconveniently large for a vacuum vessel which must be perfectly leaktight. In addition, it is desirable to remove the wall boundary layers before the flow is expanded. In the most obvious method of operation the central core of the unexpanded flow is passed into the nozzle and the rest is by-passed. In order to reduce the time taken to start the flow in the nozzle, the latter is evacuated to the lowest convenient pressure and separated from the parallel section of the tube by a light diaphragm which is ruptured by the shock wave. In this method of operation, the running time of the shock tunnel is equal to that of the simple shock tube minus the time needed to establish steady flow in the nozzle after the shock wave passes.

Reflected-shock Tunnels

If instead of by-passing the unwanted part of the stream past the nozzle the end of the tube is blanked off leaving only a small throat leading to the divergent nozzle (Fig. 15), the incident shock is reflected from the blanked-off end and travels back towards the high-pressure chamber, further compressing and heating the gas. The speed of flow in the gas traversed by the reflected shock is very small: the reflected shock quickly becomes uniform and the disturbance caused by the reflection of an expansion wave from the nozzle orifice is rapidly absorbed in the nozzle starting process. The reflected-shock tunnel is widely used for heat-transfer and pressure measurements on a variety of models: in principle it is an intermittent blow-down wind tunnel.

"Tailored" Operation

When the reflected shock hits the contact surface it is again reflected, either as a shock or as an expansion depending on the gas properties and the shock Mach number. At and near a certain Mach number the disturbance reflected from the contact surface is negligibly weak: the driven gas remains undisturbed in the near-stationary state to which it was brought by the return of the shock after its reflection from the almost-closed end of the tube, and the flow through the nozzle into the evacuated chamber beyond remains uniform for a much longer time than in the conventional shock tube or the by-pass shock tunnel, lasting until all the driven gas has escaped or until the arrival of the expansion wave which starts by progressing into the driver chamber from the main diaphragm and, after reflection, follows the contact surface down the tube. This is known as tailored-interface operation: the reflections from the contact surface are adequately weak over a range of shock Mach numbers either side of the tailoring value (theoretically 6·03 for hydrogen into air) to permit measurements to be made after a delay sufficient for the reflected disturbances to be dissipated by further multiple reflections. This is known as "equilibrium–interface" operation.

The increased running time of reflected-shock tunnels not too far from tailoring facilitates force measurements with strain gauge balances (which are affected by model vibration set up during the nozzle starting process and taking some time to die out) and static pressure gauges (which suffer from lag and resonance in their connecting tubes, however short). Stagnation temperatures as high as 12,000°K with flow Mach numbers of 20 or more can be obtained, but the static pressure in the working section, though higher than in the by-pass shock tunnel because of the compression caused by the reflected shock (a pressure ratio of the order of 8) is still much lower than in a simple shock tube with the same driving pressure. The remaining technique which can be used to improve performance is to increase the driver pressure by combustion, but the final limitation is still that of mechanical strength of the driver chamber.

The problems of shock tunnel operation cover aerodynamics, electronics and both high-pressure and vacuum plumbing. The actual measurement techniques are specialized because of the short duration of the steady flow: neither the steady-flow methods used in conventional wind tunnel testing nor the methods used for "unsteady" flow over oscillating aerofoils (which usually execute periodic motions so that time averages can be established) can be applied directly. The usual procedure is to use transducers to convert temperatures, pressures and forces into electrical signals which are then displayed on a calibrated oscilloscope and photographed for analysis at leisure. Resistance thermometers, piezoelectric crystal pressure transducers and strain-gauge balances are used: the instruments must respond rapidly to changes in the flow without overshooting the true output signal and oscillating about a mean ("ringing"). They must also be capable of withstanding large changes in input (the quantity to be measured) as the shock wave passes and in the confused flow after the passage of the contact surface.

One of the most widely used of the techniques especially developed for shock tube and shock tunnel work is the use of platinum film resistance thermometers for heat transfer measurements: heat transfer to possible re-entry shapes is of particular interest. A thin film of platinum is deposited on a glass surface: because of the large thermal capacity of the glass, the electrical resistance of the film does not change in proportion to the temperature of the gas, but the rate of increase of film temperature can be related to the rate of heat transfer from the gas by using theoretical results for unsteady heat transfer into a semi-infinite solid. Thin film thermometers are also used as shock arrival indicators: the signal from one film mounted on the tube wall is used to start an electronic chronometer as the shock passes, and the signal from a second film further down the tube is used to stop the chronometer. The shock speed and Mach number can be deduced, and the other properties of the flow such as the static pressure and flow Mach number can be calculated.

Some success has been achieved with force measurements in

shock tunnels made by observing the acceleration of a freely suspended model to avoid the vibration difficulties attendant on the use of strain-gauge balances. The model can be fitted with one or more accelerometers, or photographed at short intervals so that the accelerations can be deduced by double differentiation of distance–time records.

Other variants of the shock tube include the gun tunnel, in which the driver and driven gases are separated by a light, free piston. The gas is expanded through a nozzle as in the shock tunnel, and the sequence of operation is rather similar to that in an equilibrium-interface reflected-shock tunnel except that the disturbances are reflected from the piston face rather than the contact surface. In the hot-shot tunnel, the nozzle is supplied by highly ionized gas heated by an electrical discharge: this and several related devices are used for experiments on magnetogasdynamics, as the study of the interaction of a conducting gas with a magnetic field is popularly called.

Low-density Tunnels[25]

Shock tunnels are used primarily for measurements at high Mach number and stagnation temperature, implying a fairly high Reynolds number. In certain cases of high altitude flight, usually of lifting bodies rather than missiles, a low hypersonic Mach number is combined with a very low air density implying a low Reynolds number without an excessively high stagnation temperature. Heat transfer is therefore not of foremost importance, and there is no need to duplicate the flight stagnation temperature in model tests. Tests at very low density in short-running-time facilities are complicated by lag in leads (p. 108) and outgassing, and continuous-running low-density tunnels are more convenient to use (Fig. 16). The two most important problems are the provision of a suitable pumping system with a very high capacity by laboratory vacuum-pumping standards, and the avoidance of excessive boundary-layer growth in the convergent–divergent nozzle preceding the test section. Fans are not regarded as

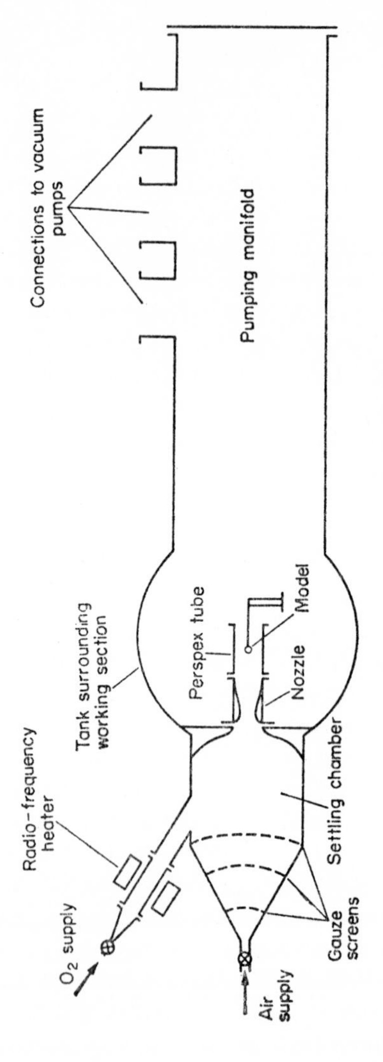

FIG. 16. N.P.L. low-density tunnel.

practicable for driving low-density tunnels, and the choice falls between ejectors and various types of vacuum pump. Ejector pumps (which are the same thing as injector pumps, but regarded as suction rather than compression devices) can be driven either by compressed air or steam, the latter being more easily generated. Conventional mechanical or mercury-diffusion vacuum pumps are adequate for small tunnels; an alternative technique is the cryo-pump in which the gas to be pumped is liquefied or solidified in contact with a surface cooled on its other side by liquid helium. Low-density tunnels are usually operated at Mach numbers of 5 to 10, requiring a very large expansion ratio in the nozzle; the nozzle boundary layers grow so quickly at low Reynolds numbers that the usable core of uniform flow in the working section is sometimes very small. The experimental techniques are complicated by the small absolute values of the pressures and forces to be measured: it is usual to install the manometers as well as the force balance inside the evacuated shell of the tunnel, and the leads to the manometer are kept as short as possible to minimize lag and out-gassing effects.

Low-density flows can also be used in the study of fluidic (fluid logic) devices, which are normally too small for detailed flow measurements to be made; the right Reynolds number can be obtained with a model n-times full scale operated at $1/n$ of normal density.

Examples

1. Suggest a rapid method of finding whether the fan of a wind tunnel is operating near its conditions of maximum efficiency.
2. Why could not the screens and honeycombs in a test rig not be combined in a honeycomb with cells of very small diameter, producing a pressure drop coefficient of the same order as a screen?
3. How could the high "turbulence" intensity in supersonic tunnels (see p. 66) be reduced?
4. Why are shock tunnel nozzles usually axisymmetric while wind tunnel nozzles are usually two-dimensional?

CHAPTER 3

Fluid Velocity and Shear Stress Measurements

Pitot Tubes

As mentioned on p. 19, the fluid velocity at a point can be found by measuring the total pressure and the static pressure, and applying eqn. (4) or (6). The total pressure can be measured as the pressure at the front stagnation point of any suitable body: an obvious shape of body to choose is a tube aligned along the flow direction, known as a Pitot tube after its inventor. The pressure recorded by the tube is closely equal to the total pressure, to an accuracy of, say, $\frac{1}{4}$ per cent of the dynamic pressure, provided (i) that the Reynolds number based on tube diameter is more than about 100, (ii) that the tube is aligned within about ± 10 deg of the flow direction, (iii) that the root-mean-square intensity of turbulence in the stream is less than about 5 per cent of the mean velocity, (iv) that the total pressure does not change by more than 1 to 2 per cent across the tube diameter, and (v) that the probe is not too near a wall. In addition we must remember that in supersonic flow the pressure recorded by the tube will be the total pressure behind a normal shock wave. Naturally the strict observance of all these conditions would leave very few situations in which a Pitot tube could be used, and most of the information required for the interpretation of Pitot tube measurements consists of a description of the various corrections to be applied to the readings for the effect of low Reynolds number, yaw, high turbulence, transverse total-pressure gradient and wall proximity. A thorough review of the Pitot tube[26] occupied 92 pages and quoted 147 references, so it will be seen that the use of Pitot tubes is not as straightforward as would appear from eqns. (4) and (6). The Pitot tube, associated with a probe for static pressure measurement, is the most widely used instrument for fluid velocity measurement, and can also be adapted for direction or surface shear stress measurement, but other methods are necessary for measurements of violently fluctuating flows and of the rate of flow in ducts[27]. The problem of measurement of surface shear stress, that is of the distribution of the drag force exerted by viscous effects at the surface, can be reduced to the estimation of the velocity gradient at the surface, but it is possible to do this directly only in special cases, and a number of indirect methods are currently in use.

Corrections to the Readings of Pitot Tubes in Non-uniform Flow

(*a*) *For Total-pressure Gradient.* Experiments performed about 30 years ago showed that the pressure in a cylindrical Pitot tube in a shear layer was equal to the total pressure at a point displaced about 0·15 diameters from the centre of the tube in the direction of increasing velocity, and more recent measurements[28] have confirmed this correction for steady subsonic flow. By far the majority of Pitot tube measurements are made in boundary layers or other thin shear layers (the total pressure in most flows being constant outside such layers) and for accurate work it is customary to use flattened tubes (Fig. 17) to minimize the uncertainty both of this displacement effect and of the further displacement of the "effective centre" caused by the proximity of a wall. Tubes with tips

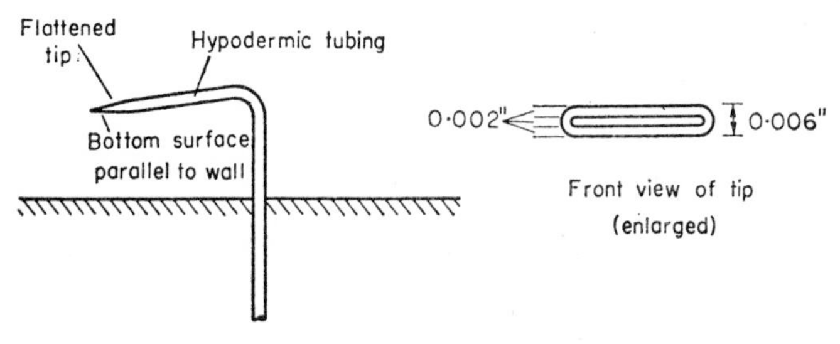

FIG. 17. Flat Pitot tube for shear layer measurements.

0·015 cm high can be made quite easily from hypodermic tubing. Flat tubes seem to be very sensitive to angle of incidence and are not suitable for use in highly turbulent flow. Very small round tapered tubes have been made by drawing quartz tubing, but the flat tube has the advantage of a larger hole area for a given effective height, so that lag due to the volume of the leads and the manometer (see p. 108) is reduced. Burrs and imperfections of microscopic size on these fine tubes can affect the pressure coefficient $(p_0' - p)/(p_0 - p)$ where p_0 and p_0' are the true and measured total pressure respectively, and p is the static pressure: for a perfect tube this coefficient would be unity. It is easier to calibrate small tubes individually than to take especial care in manufacture.

(*b*) *For Proximity to a Surface.* The correction of Pitot tube readings for proximity to a solid surface is less straightforward than the correction for transverse velocity gradient because the error depends on the whole of the velocity profile between the tube position and the wall. Very possibly a successive-approximation procedure could be developed to deduce the true profile from the measured profile but a great deal of experimental work would be needed to follow such a procedure. As indicated above the practice is to use very small tubes for measurements in thin boundary layers and to ignore the correction entirely. According to reference 28 the correction to the velocity measured by a round Pitot tube is less than $\frac{1}{4}$ per cent of the dynamic pressure $(p_0 - p)$ if the centre line of the probe is more than about one diameter from the surface: the correction for a flat tube may be expected to be greater because the effect of the wall is to prevent the streamlines of the flow from being deviated by the presence of the tube so that the greater width of the flat tube will confine the streamlines to a greater extent than the round tube.

(*c*) *For Turbulence.* In turbulent flow, the reading of a round Pitot tube is, it is generally agreed, increased by a quantity of the order of $\frac{1}{2}\rho(\overline{u^2} + \overline{v^2} + \overline{w^2})$ if the three mean-square fluctuation components are roughly equal. The exact error in total pressure measurement depends on the distribution of turbulent energy among eddies of different sizes relative to the size of the tube, and on the dynamic response of the tube-lead-manometer system. If the fluctuations in the oncoming stream are sufficiently slow for the flow around the tube, or any other body, to be the same at any instant as if the stream were steady with the same speed and direction, the flow is said to be "quasi-steady". As an example, the flow around a bluff body, like the elliptic cylinder shown in Plate 5, would be quasi-steady if the stream fluctuation frequency were much less than the frequency at which vortices are shed into the wake. In the particular case of a circular cylinder transverse to the stream the vortex shedding frequency is approximately $0 \cdot 2U/d$ Hz, so that the flow would be quasi-steady if the free-stream fluctuation frequencies were less than $0 \cdot 02U/d$, say. The concept of quasi-steadiness is helpful in many problems of unsteady motion of a stream relative to a solid body: in this case it enables the time-mean total pressure to be calculated from the indicated total pressure if the variation of the pressure coefficient of the Pitot tube with yaw angle is known. If we assume that the flow is quasi-steady and that the lateral component of turbulence intensity is low enough for the effect of yaw angle to be negligible, and put V, the resultant velocity, equal to the vector sum of the mean velocity U and the fluctuating components u, v and w, then we see that the tube reads $p + \frac{1}{2}\rho\overline{V^2} = p + \frac{1}{2}\rho U^2 + \frac{1}{2}\rho(\overline{u^2} + \overline{v^2} + \overline{w^2})$. If, on the other hand, the velocity is constant in magnitude but varies in direction over a wide angle, the tube will be expected to read *low* because the recorded pressure is smaller than the true total pressure at large angles of yaw. Needless to say, it is very doubtful whether the quasi-steadiness assumption is even approximately valid for thin turbulent shear layers in which the eddies may extend down to a range of wavelengths much less than the length l of the tube so that $fl/U \gg 1$. The present position of our knowledge is highly unsatisfactory, particularly as regards the use of Pitot tubes in highly turbulent jets, but it is difficult to believe that any simple formula can ever give the correction to be applied to measurements in a general turbulent flow without extensive data on the turbulence itself: if turbulence measurements are required as primary data it will be necessary to make them with a hot wire anemometer, whose output voltage can be made a linear function of velocity so as to indicate true mean velocity directly, avoiding the use of a Pitot tube altogether.

(*d*) *For Yaw*. The effect of mean yaw or incidence on the reading of round tubes with plane ends is small for angles less than about 10 deg: it seems to be rather larger for hemispherical- or ellipsoidal-ended tubes, and flat Pitot tubes are even more sensitive to incidence. In nominally two-dimensional flow the incidence can usually be guessed to within a few degrees and the error is negligible: in three-dimensional flow the flow direction is usually required anyway and has to be measured with a yawmeter.

Static Pressure Measurements

If a hole is made in the side of a forward-facing tube instead of at the front end, the tube will indicate the static pressure approximately. Unless the tube is infinitely slender and the hole infinitely small (see pp. 106–7) the reading will not be exactly equal to the true static pressure but will differ from it by a fraction of the dynamic pressure which depends on the shape of the tube and on the Mach and Reynolds numbers. The pressure coefficient will also depend on the yaw angle and the turbulence in the stream,[29] but the effect of a transverse total-pressure gradient is usually quite negligible.

The N.P.L. standard Pitot-static tube (Fig. 18) combines a hemispherical-ended Pitot tube with a static tube in which the holes are placed so that the pressure coefficient of the static tube is nearly zero for incompressible flow at all reasonably high Reynolds numbers, the pressure disturbances induced by the nose shape and by the perpendicular support tube cancelling out. This tube is 0·307 in. (0·78 cm) in diameter and Reynolds number effects are negligible at normal speeds, but the smaller tubes used for shear-layer exploration may suffer slight Reynolds number effects at conventional working pressures and the effects may be serious in low-density flow. As usual, the safe procedure is to calibrate the tube in the Reynolds number range in which it is to be used.

Static pressure tubes of the conventional type are more sensitive to yaw than Pitot tubes: in subsonic flow the pressure coefficient $(p' - p)/(p_0 - p)$ of the static-tube portion of the N.P.L. standard instrument is about $2(1 - \cos \alpha)$ where α is the yaw angle. The reading in a turbulent flow may be either higher than the true static pressure (for a large tube) or lower (for a small one). Gadd[30] has described a wedge-shaped static tube, in appearance like a flat

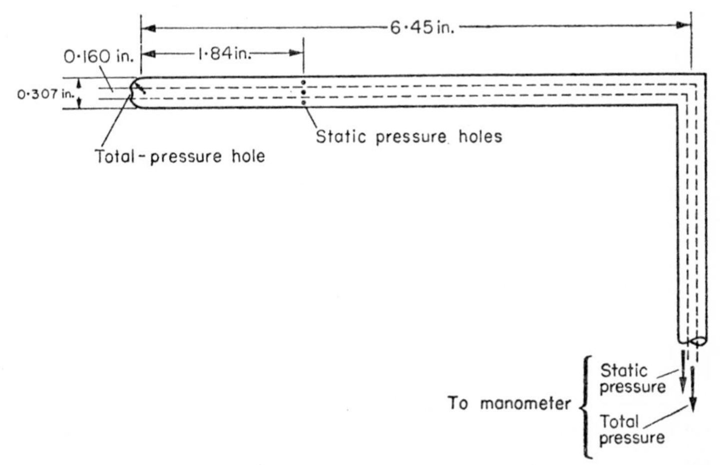

FIG. 18. N.P.L. standard Pitot-static tube.

Pitot tube with the hole in the side instead of the front, which was first used in France for supersonic flow measurements.

The static pressure in a shear flow for which the boundary-layer approximation is valid is equal to the pressure on either side of the shear layer, or, in the case of a turbulent flow, to this pressure diminished by $\rho \overline{v^2}$, as can be seen from the equation of motion for the y-component (see p. 28). In the case of boundary layers it is most convenient to measure the pressure on the surface. If the shear layer is appreciably curved (like the boundary layer on a thick aerofoil or the jet of a jet flap wing) it is probably more accurate, if more tedious, to calculate the static pressure from the measured total pressure and the radius of curvature of the flow (see example 2). The reading of a static tube will clearly be affected if the radius of curvature of the flow is not very large compared with the length of the tube. Pitot-static tubes are not suitable for detailed flow explorations or for use in restricted spaces because the total pressure and static pressure are measured at different points and because the narrow-bore tubing needed for the inner of the two concentric tubes (see Fig. 18) increases the lag considerably if the probe is made small enough for the purpose. Any body with two pressure tappings can in principle be used to measure total and static pressure by solving the equations

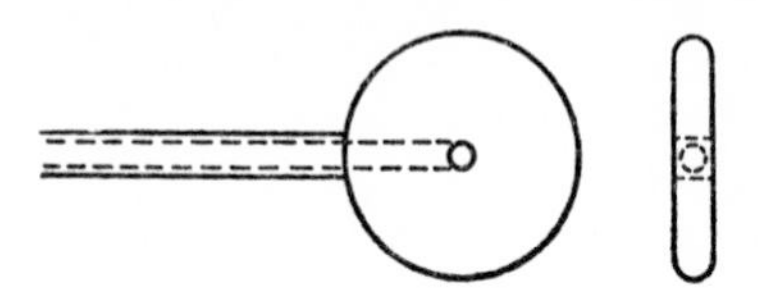

FIG. 19. Disc static probe (with flow in plane of disc).

$p_1 = p + k_1(p_0 - p)$ and $p_2 = p + k_2(p_0 - p)$, where p_1 and p_2 are the pressures measured at the two holes, and k_1 and k_2 are constants obtained by calibration in a known flow. Several probes employing this principle have been developed for special purposes.

An ingenious instrument for use in three-dimensional flows is the disc static probe[31], Fig. 19. The holes on either side of the disc are connected to a *single* tube leading to the manometer, with the result that the reading of the probe, as well as being completely independent of flow angle in the plane of the disc, is almost constant for angles of inclination less than about ± 5 deg in the plane perpendicular to this. At larger angles of yaw the error increases rapidly, but in most three-dimensional flows one can guess the flow angle to sufficient accuracy in the plane in which it varies less, and mount the disc perpendicular to this plane. The disc static probe has a pressure coefficient at zero yaw of about $-0 \cdot 1$ and individual calibration in the Reynolds number range of interest is advisable.

Fluctuating Pressure Measurements

There is not a great deal of interest in fluctuating total pressure, except in internal combustion engines and in establishing the pressure indicated by a Pitot tube in turbulent flow. In principle, a small pressure transducer mounted at the front of a Pitot tube will indicate the instantaneous total pressure, at least as long as the flow over the tube can be regarded as quasi-steady and if yaw effects can be disregarded. The measurement of fluctuating static pressure, on the other hand, is of the greatest interest in investigations concerning buffeting of bodies by separated flow, and in aerodynamic noise. The measurement of fluctuating surface pressure with flush pressure transducers is aerodynamically trivial: the mechanical problems and the problems of pressure gauges some distance below the surface are discussed in the next chapter. The measurement of fluctuating static pressure within a turbulent flow is at present an unsolved problem: one requires a probe whose reading is independent of dynamic pressure, and insensitive to yaw in any direction through angles of at least 15 deg.

Flow Direction Measurements

Pressure holes symmetrically placed on either side of a body will show different readings if the body is yawed. The flow direction can be found either by calibrating the body to find the variation of the pressure-difference coefficient with yaw angle, or by rotating the body until the pressure difference is zero. In general the latter, null-displacement, method is the more accurate, although it is still wise to make a preliminary test in a stream of known direction to allow for any slight differences in hole position between the two sides: the main disadvantage of the null-displacement method is that the probe attitude must be adjusted by remote control, which is particularly difficult if the probe is also to be traversed linearly across the flow, or if both yaw angle and incidence are to be measured.

Fixed-direction yawmeters (Fig. 20), which indicate flow angle in terms of a pressure difference, usually incorporate a total-pressure hole as well as two or four yaw holes depending on whether the flow to be explored is two- or three-dimensional. Over a limited range of yaw angles, the average pressure coefficient

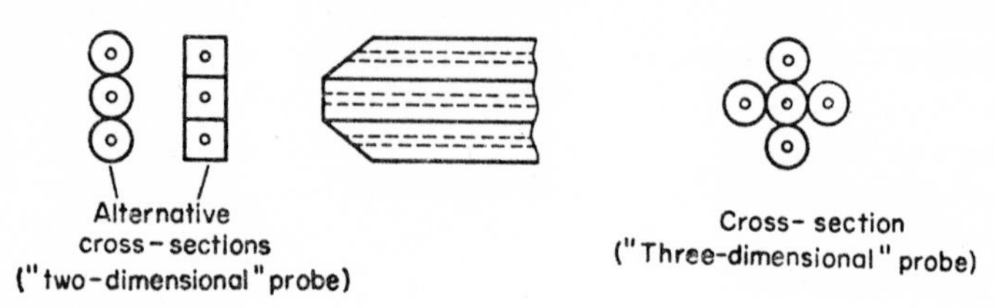

FIG. 20. Yawmeter.

$(p' - p)/(p_0 - p)$ of the yaw holes will be equal to that at zero yaw, and so the static pressure and dynamic pressure can be deduced from the average pressure of the yaw holes and the total pressure measured by the hole at the front of the tube. For larger yaw angles the pressure coefficient of a yaw hole will be a noticeably non-linear function of yaw angle, and the effect of yaw on the reading of the total pressure hole may also have to be taken into account, so that it is wise to align the probe in an estimated mean flow direction before starting a traverse in order to minimize the relative yaw angles. A further inaccuracy in the use of small yawmeters for detailed exploration of three-dimensional shear layers arises from their liability to appreciable Reynolds number effects. Reference 31 gives a full discussion of yawmeters and static probes for three-dimensional flow measurements.

Flow Quantity Measurements

The average velocity in a duct can be found from measurements of the velocity over the whole of the cross-section by traversing

Pitot and static tubes, but this is tedious and there are several devices which yield the average velocity in a circular pipe from two pressure readings to an accuracy sufficient for most purposes, and whose calibration is well documented.

The orifice-plate flowmeter (Fig. 21(a)) is the crudest, and probably the most widely used in cases where the pressure drop caused by the flowmeter is unimportant. The pipe is obstructed by a plate with a circular, sharp-edged hole, and the static pressure on the wall of the pipe is measured at points just upstream and just downstream of the plate or, to simplify construction, at one pipe

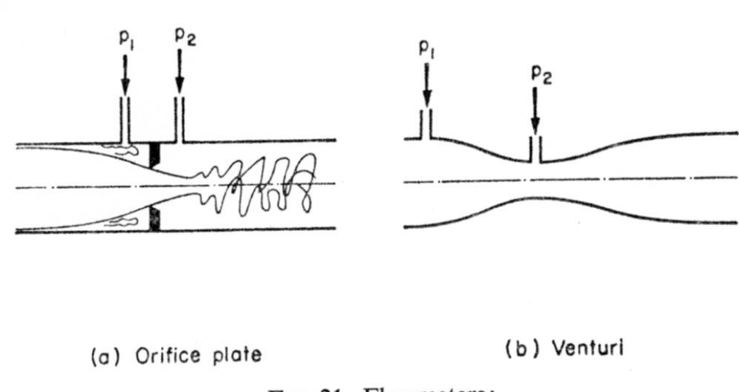

FIG. 21. Flowmeters:
(a) Orifice plate.
(b) Venturi.

diameter upstream and half a pipe diameter downstream of the plate. The average velocity is specified as an empirical fraction, α, of the velocity calculated on the assumption that the pressure difference is equal to the difference in dynamic pressure between the flow through the orifice and the flow through the pipe, based on average velocities in each case. This amounts to the assumptions that the flow separates from the walls of the pipe in the adverse pressure gradient which necessarily precedes the acceleration of the flow through the crude contraction formed by the orifice (see p. 57) and that the static pressure in the separated region is equal to the static pressure which would exist if the orifice plate were not there: it is further assumed that the static pressure just downstream of the plate is constant over the whole of the cross-section. Naturally, in view of the crudity of these assumptions, the discharge coefficient α is not very near unity: it takes values of about 0·6 for orifice diameters from 0·2 to 0·6 of the pipe diameter and for Reynolds numbers, based on the orifice diameter and the velocity through the orifice, of more than 10^4. The chief reason for the large departure of the discharge coefficient from unity is that the flow continues to accelerate downstream of the sharp-edged orifice, forming a "vena contracta" whose cross-sectional area is less than that of the orifice, and whose static pressure, which determines the static pressure at the wall, is therefore less than the static pressure corresponding to the assumed uniform flow through the orifice. Discharge coefficients much nearer unity can be achieved by using a shaped nozzle so that the flow is more nearly uniform and parallel at the orifice, but construction is more expensive. The advantages of the orifice-plate flowmeter are that it is robust and that calibrations are easily reproduced: the variation of α over a large range of diameter ratio and Reynolds number is given as Fig. 133 of ref. 8 (see also ref. 32). Since the variations of α with Reynolds number are caused chiefly by changes in the velocity profile of the undisturbed flow in the pipe, it would be more logical to use a Reynolds number based on pipe diameter and mean velocity, but of course the two Reynolds numbers are uniquely related by the diameter ratio. The orifice plate flowmeter is reasonably insensitive to Reynolds number and to asymmetry and perturbations in the pipe flow caused by bends and obstructions upstream: the discharge coefficient of an orifice plate with a diameter ratio of 0·5 is changed by only about $\frac{1}{2}$ per cent by a right-angled bend 5 diameters upstream.

A more efficient device is the Venturi flowmeter (Fig. 21(b)) which again consists of a contraction in area, with a smooth entry like the shaped nozzle, but followed by a gentle diffuser so that the total-pressure losses scarcely exceed those in an equal length of straight pipe. The static pressure is measured at the entry to the contraction and at the throat, and the discharge coefficient, defined as before, is between 0·95 and 0·98 depending on Reynolds number. The orifice plate is preferred for reliability and ease of construction whenever the pressure drop caused by the flowmeter is unimportant. A flow meter with a fairly low pressure loss and almost complete independence of Reynolds number and upstream disturbances can be made from a diffuser, honeycomb and screen, and a contraction back to the original diameter in which the flow is uniform except for fairly thin boundary layers on the walls: the velocity is deduced from static pressure tappings on the contraction walls in exactly the same way as for a wind tunnel[33]. A choked (sonic) nozzle can be used for gas flow measurements[34] if a large pressure drop can be accepted. The mass flow is theoretically

$$\rho_0 a_0 A \left(\frac{2}{\gamma+1}\right)^{\frac{\gamma+1}{2(\gamma-1)}}$$

where ρ_0 and a_0 are the density and speed of sound at stagnation conditions, inferred from one total-pressure and one total-temperature measurement: in practice a discharge coefficient must be used, as in low-speed flow.

If the flow in the pipe is turbulent and fully developed, as is usually the case in industrial pipelines carrying fluid of reasonably low viscosity, the average velocity can be found from a measurement of the velocity on the centre line and a knowledge of the ratio of centre line velocity to mean velocity obtained from the measurements of previous workers. This type of flowmeter would not disturb the flow appreciably but it would be very sensitive to asymmetry or perturbation of the flow and rather sensitive to surface roughness. Preston (see ref. 8) has developed this simple idea by using four Pitot tubes at 90 deg intervals on a circle whose radius is three-quarters of the pipe radius, together with four static holes. This radius was chosen as being that at which the velocity is a constant fraction of the average velocity over a wide Reynolds number (and roughness) range. The "three-quarter radius" flowmeter requires a settling length of at least ten diameters, but is a useful and simple instrument for measuring the rate of flow through a duct connected to a pumping system of limited power.

Mass flows can be measured by introducing a known (small) mass flow of another fluid, or a solute, or a known heat source, and then measuring concentration or temperature rise far enough downstream for mixing to be complete. The potential difference developed in a conducting liquid flowing through a magnetic field can also be used.[35]

Fluctuating Velocity Measurements—The Hot Wire Anemometer

As we have seen, Pitot and static tubes can barely be relied upon to indicate the *mean* velocity of an unsteady flow with sufficient precision, and their use for measurements of the fluctuations themselves can rarely be considered. Nearly all fluctuation measurements are made with hot wire anemometers[36] or other heated bodies. The temperature attained by a heated body in an airstream depends on the velocity of the stream, so that the body can be calibrated to indicate velocity in terms of temperature or rate of heat transfer. Dimensional analysis, for which there is great scope in heat transfer, shows that the heat transfer coefficient, also called the Nusselt number, is a function of the Reynolds number if the heat transferred by free convection is negligible. The Mach number, the ratio of specific heats γ, and the Prandtl number $\mu c_p/k$ (where c_p is the specific heat at constant pressure and k the thermal conductivity of the fluid) also influence the heat transfer: the Prandtl number is a constant for a given perfect gas, though its variation with temperature in liquid flows may be important. The Nusselt number is

$$N_u = \frac{\text{rate of heat transfer}}{k \times \text{temperature gradient} \times \text{area}}$$

If we consider the electrical heating of a cylindrical wire normal to a fluid stream we find

$$N_u = \frac{I^2R}{k(T - T_a)(1/d)\,\pi dl} = \frac{I^2RR_a a}{\pi lk\,(R - R_a)} \qquad (19)$$

where I is the electrical heating current and R the resistance, l the length and d the diameter of the wire, T the temperature of the wire (assumed uniform over the length and cross-section) and T_a the temperature of the ambient fluid. α is the temperature coefficient of resistance, here defined by $R = R_a[1 + \alpha(T - T_a)]$ using the ambient temperature rather than 0°C as the reference temperature, and taking an average value of temperature coefficient between the temperature of the fluid and that of the wire. α may vary slightly with either temperature and k varies appreciably with temperature: since we wish to non-dimensionalize the heat transfer across the temperature difference $T - T_a$ we ought to insert some average value of k in the formula for N_u, and the usual procedure is to evaluate it at $(T + T_a)/2$, the mean of the wire and fluid temperatures. This is only an approximation, and the correctly weighted value of k could only be estimated if the temperature distribution throughout the flow were known theoretically or experimentally, but quite often the variation of k and α is neglected altogether.

Hot Wire Probes

In its practical form the hot wire anemometer consists of a thin ($d < 0{\cdot}05$ mm) wire not more than 0·5 cm long. The upper limit of length is set by the spatial resolution required, and the diameter is limited by the need for an adequate electrical resistance and by the need for the wire to respond to rapid fluctuations in heat transfer due to changes in velocity. The usual materials are platinum or platinum alloy, which can be obtained as a thin core surrounded by a silver sheath (Wollaston wire) to produce a wire thick enough for easy handling, and tungsten, which is produced in small diameters for electric lamp filaments and is used when greater strength is required. The disadvantage of tungsten is that it cannot be soft-soldered: Wollaston wire can easily be soldered to the supporting prongs of the probe (Fig. 22) and the silver sheath then etched away with acid except in the vicinity of the soldered joints: it is often convenient to copper-plate tungsten

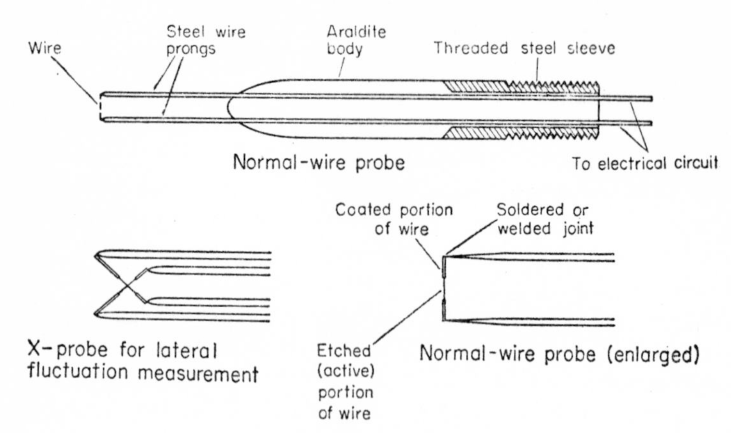

FIG. 22. Hot wire probes.

and treat it like Wollaston wire, but otherwise it must be welded to the prongs which is a difficult operation with very fine wires, although commercial probes are made in this way.

The wire can be operated either at constant current, by using a battery with a large series resistance, or at constant resistance (i.e. constant temperature) as one arm of a Wheatstone bridge (Fig. 23). The determination of the Nusselt number as a function of Reynolds number can then be simplified to finding the variation with U/ν of $R/k(R - R_a)$ or I^2/k for constant-current or constant-temperature operation respectively, since the quantities omitted from the Reynolds number and Nusselt number are constant for a given wire. The velocity fluctuation intensity can then be deduced from

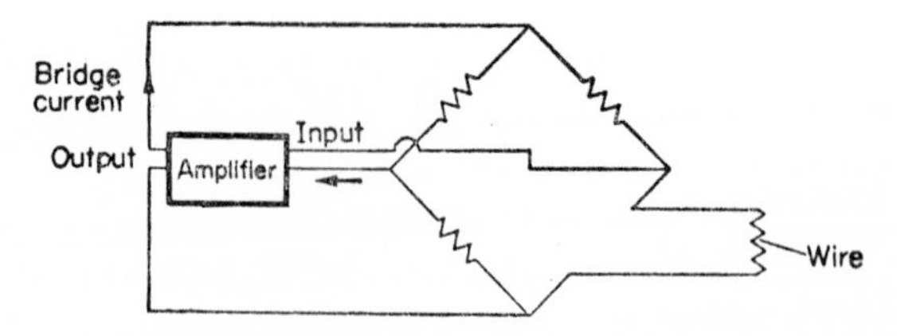

FIG. 23. A Wheatstone bridge for constant-resistance operation.

the voltage fluctuations by dividing the latter by $I\ \partial R/\partial U$ or $R\ \partial I/\partial U$ respectively. The temperature ratio T/T_a, roughly equal to R/R_a, is kept above 1·3 to minimize errors due to small changes in fluid temperature—since a hot wire is merely a resistance thermometer with a rather large excitation current—and below 2·0 to avoid the effect of large changes in α and k, and of oxidation or weakening of the wire, at higher temperatures. The temperature ratios used in water and other liquids must be much smaller to avoid bubbles of dissolved gas collecting on the wire (*cooled* probes have been used), and wires used in conducting liquids should be electrically insulated by a thin, vacuum-deposited film of quartz.

A wire inclined at an angle to the flow will respond to the component of the velocity normal to the wire: this is a rough rule, considerably modified by the effects of finite wire length, so that if inclined wires are to be used for measurement of v or w component fluctuations it is usual to calibrate their response to yaw as well as to variations in stream velocity. The difference between the outputs of two identical wires arranged in an X (Fig. 22) can be seen to be a function of the lateral velocity component (i.e. the component up or down the page in Fig. 22, the main flow being from the left).

The hot wire is sensitive to dust in air and slime in water, so that frequent recalibration is necessary: some advantage in stability and strength can be gained by using a platinum film on the front of a glass wedge, but the hot-film probe is much more difficult to manufacture and its response cannot be extended to such high frequencies as that of a wire. Also, conduction of heat to the fluid via the glass affects the low-frequency and steady-state calibration[37]: this effect is small in liquids because the thermal conductivity of the glass is small compared with that of the liquid, but the same is not true in gases. Neither the hot film nor the hot wire would be used for mean velocity measurements unless a Pitot tube was quite unsuitable, and their chief use is for fluctuation measurements for which they are virtually unequalled: any method which used a less temperamental sensing element would be welcomed. Occasionally, as in the study of fluidic devices, their small size is an advantage.

Compensation for Thermal Inertia of Hot Wires

Although the thinnest wires (wires 0·00025 cm in diameter and 0·5 mm long have been used in supersonic tunnels) will follow velocity fluctuations up to several hundred Hertz, this frequency response is often inadequate for measurements in fully developed turbulence. The limit is set by the thermal inertia of the wire: the time taken to transfer a quantity of heat equal to that required to change the wire temperature by a finite amount is not infinitesimal. The differential equation relating the wire temperature to time is linear and of the first order, and it follows from this that the response to a small step change in velocity would be $T = T_1 + \Delta T[1 - \exp(-t/M)]$ where T_1 is the initial temperature and $T_1 + \Delta T$ the final temperature of the wire. M is a quantity having the dimensions of time and called the time constant of the wire: it is inversely proportional to the rate of heat transfer and is therefore a function of the operating conditions. If a sinusoidal velocity fluctuation of frequency f is applied to the wire it again follows from the differential equation that the corresponding temperature fluctuation will be a fraction $1/[1 + (2\pi fM)^2]^{1/2}$ of the amplitude predicted from the mean-speed calibration of the wire. If as is usual when dealing with the frequency response of electronic apparatus we define the maximum useful frequency (the upper limit of quasi-steadiness of the temperature of the wire) as that at which the fluctuation is $1/\sqrt{2}$ of its low-frequency value we see that this frequency is $1/2\pi M$ Hz.

The differential equation governing the change of wire temperature with time for a given velocity fluctuation is the same as that governing the change of output voltage with time for a given input voltage fluctuation for the resistance–capacitance electrical filter circuit shown in Fig. 24, a so-called low-pass filter, if $RC = M$. In order to compensate for thermal inertia we can pass the electrical signal from the wire through a circuit which amplifies it by a factor which *increases* with frequency as $[1 + (2\pi fM)^2]^{1/2}$. The easiest circuit to understand is shown in Fig. 25(a). If we use

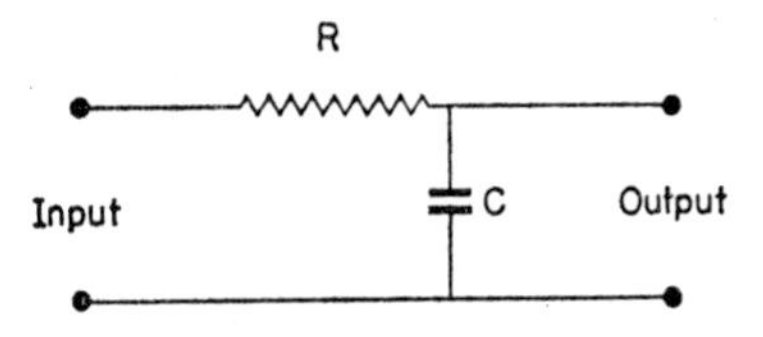

FIG. 24. Low-pass electrical filter.

complex notation we can write the required amplification as $1 + j\omega M$ where $j = \sqrt{-1}$ and $\omega = 2\pi f$, corresponding to the wire response $1/(1 + j\omega M)$. The actual response of the resistance–capacitance network of Fig. 25(a) is

$$e_0/e_i = \frac{R_2(1 + j\omega R_1 C)}{R_1 + R_2(1 + j\omega R_1 C)} \tag{20}$$

where e_i and e_0 are the input and output amplitudes of the sinusoidal fluctuation of frequency $\omega/2\pi$. If $R_2 \ll R_1$ so that the denominator is approximately R_1 for all frequencies of interest we have $e_0/e_i = (R_2/R_1)(1 + j\omega R_1 C)$ so that the circuit will compensate the wire output e_i for thermal inertia if it is adjusted so that $R_1 C = M$. The circuit attenuates the wire signal by the factor

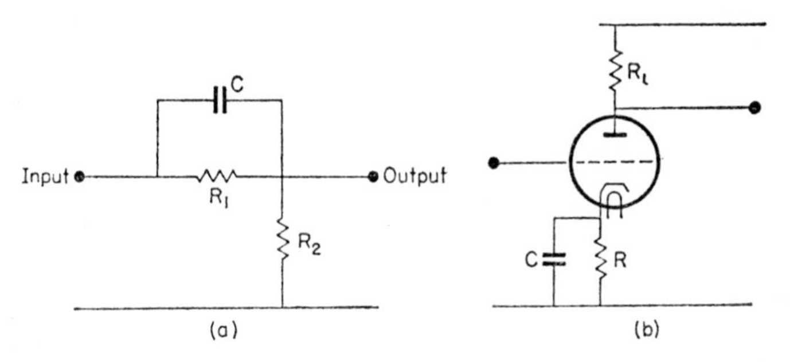

FIG. 25. Hot wire compensation circuits.

R_2/R_1 at low frequencies, and this attenuation is by definition large, so that it would be necessary to pass the signal through a valve amplifier to restore it to its original level.

A more acceptable circuit in which, crudely speaking, the filter network and amplifier are combined into one, is shown in Fig. 25(b). Here a resistor and capacitor are placed in parallel in the cathode circuit of a valve or the emitter circuit of a transistor. At low frequencies the resistor "feeds back" most of the output signal to the grid in opposition to the input signal, and the net amplification of the whole circuit is approximately R_l/R which can be adjusted to unity if desired. At higher frequencies, the capacitor C shunts the resistance R, changing the complex impedance of the cathode circuit from R to $R/(1 + j\omega RC)$ so that the net amplification changes from R_l/R to $(R_l/R)(1 + j\omega RC)$ as required for wire compensation. The frequency range for proper compensation is limited by the amplification of the circuit in the absence of feedback, assumed infinite in the above simplified analysis. Second-order effects, like longitudinal heat conduction in the wire, become important at very high frequencies.

Constant-temperature Hot-wire Operation

In the above discussion of thermal inertia we have implicitly assumed that the wire heating current remained constant when the velocity changed, and that in consequence the temperature varied. It is possible to maintain the instantaneous temperature of the wire constant by supplying the heating current from an amplifier whose input depends on the departure of the wire temperature (or resistance) from the desired value: this is done by connecting the amplifier to a Wheatstone bridge circuit containing the wire as shown in Fig. 23. The bridge out-of-balance signal is amplified and fed back to the bridge as a change of wire current of the right sign to restore the wire resistance to the value for which the bridge is balanced. In this way, the effective time constant of the wire can be reduced in direct proportion to the gain of the amplifier and no further compensation is needed.

The advantage of this system over the constant-current arrangement with a compensation circuit is that no manual adjustment is needed if the wire operating conditions or the velocity of the flow change appreciably (and that in particular the wire does not burn out if the flow stops). The wire is therefore capable of reproducing large fluctuations in velocity without the uncertainties entailed by the resulting large fluctuations in time constant. The signal is still distorted because the variation of wire current is not linearly proportional to the velocity, but if the wire calibration is known it is possible to linearize the output with a special amplifier and the system is then capable of handling the very large turbulent intensities which occur in jets and separated flows. The disadvantage of the constant-temperature system is that the amplifier is more complicated. Far more constant-temperature sets are sold commercially than constant-current sets, but many people build their own constant-current apparatus.

The Doppler Anemometer

The frequency of light (or sound) scattered at angle β to the track of a particle moving at speed U differs from the frequency of the incident beam by a fraction

$$(\cos\beta - \cos\alpha)U/c,$$

where α is the angle between the particle track and the incident beam (Fig. 26) and c is the speed of light (or sound) in the surrounding fluid, *not* the speed *in vacuo*. If the frequency difference can be obtained directly, the particle velocity can be measured directly as the output of a frequency-modulation discriminator, the principle being the same as an FM radio. This technique has been developed recently, using a laser as a light source. A block diagram of a typical system is shown in Fig. 26. Doppler-shifted scattered light and an unscattered beam meet on the surface of a non-linear photo-detector, so that the output contains a term proportional to the square of the sum of the two signals, and this term contains a component at the difference frequency ("beats").

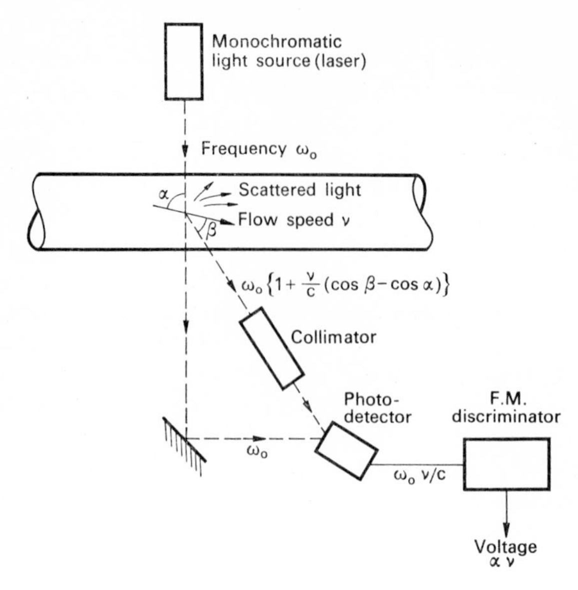

FIG. 26. Doppler anemometer.

This is a version of the well-known heterodyne technique: since two signals from the *same* source are compared, it is more properly called "homodyne". The volume of resolution is the portion of the beam that appears in the field of view of the microscope. One usually wants to make this as small as possible, but the volume of resolution must contain a large number of scattering particles (smoke or dye) and have length dimensions of many light wavelengths, because discontinuities in scattering caused by the entry or exit of the particles produce spurious outputs analogous to the "noise" produced by a hot wire anemometer. Traversing the volume of resolution through the flow presents difficulties, as does the rearrangement of the system to measure another component of velocity. These purely mechanical difficulties are being overcome and at least one self-contained laser anemometer is on the market at the time of writing.

Recording Techniques

A discussion of the recording techniques used with hot wire anemometers and other transducers for fluctuating flow measurements could occupy a book by itself. As a rule, information on turbulence is required in statistical form, describing the root-mean-square value, frequency spectrum and probability distribution of the signal and its correlation with the signal from transducers at other points in the flow. The methods used are basically those of analogue computing, squares and products of signals being generated by non-linear amplifiers or other devices, but the details of the circuits may be very different because a much higher frequency response is required than normally suffices for analogue computing. If the signal is sampled digitally (a sampling rate at least twice the highest frequency in the signal is needed to ensure proper resolution) the statistical analysis can be done on a digital computer. This technique may be expected to become more popular in the future but it scarcely seems worth while for the simpler quantities like root-mean-square values, for which analogue apparatus is readily available.

Measurement of Very Low Velocities

The smallest pressure difference that can be comfortably and accurately measured on a manometer is about 1 N m^{-2}, corresponding to the dynamic pressure of an airstream of 1 m s^{-1}. Hot wire anemometers can be used for mean velocity measurement down to the speed at which the heat lost by free convection is much larger than the heat lost by forced convection. This lower limiting speed can be reduced by using a cooler wire, but low-speed flows tend to be non-uniform in temperature unless they have been established for a very long time, and so errors may become appreciable if the hot wire temperature ratio is small. Nevertheless, speeds down to two or three cm s^{-1} can be measured with fair accuracy by special types of hot wire. The pulsed hot wire[38] is one of many devices in which velocity is deduced from the time of flight of a parcel of contaminant. A pulse of electrical power is applied to one wire and the resulting thermal wake is detected by a second, downstream wire used as a resistance thermometer. It is almost unaffected by changes in fluid temperature, but free convection effects become important, at least in vertical flows, at very low speeds. One advantage of time-of-flight anemometers is that they measure the component of velocity in the direction of the base line whereas most other anemometers have a complicated yaw response.

A clever application of the principles of dimensional analysis is the measurement of velocity by observing the frequency of vortex shedding behind a bluff body such as a circular cylinder, by using smoke or dye tracers and a stroboscope, or of course a hot wire and a frequency meter. The non-dimensional frequency or Strouhal number fd/U is known accurately over a large range of Reynolds numbers Ud/ν, but could be checked by measurements on a smaller cylinder at a speed high enough to measure by conventional techniques, to give the same Reynolds number. For instance, a speed of 30 cm s^{-1} could be measured with a cylinder 0·5 cm in diameter (giving a Reynolds number in air of 100, which is about the lowest at which the vortex street is easily detectable) for which the frequency would be about 10 Hertz. The calibration could be done at about 3 m s^{-1} with a 0·05 cm diameter wire, giving the same Reynolds number and a frequency 100 times as great. The instrument could also be used to measure fluctuating flows, providing that the fluctuation frequencies were an order of magnitude less than the vortex-shedding frequency, i.e. if the flow were quasi-steady: a change in velocity would appear as a change of frequency.

A very simple device for measuring low speeds, particularly in fluctuating flow, is the fibre anemometer, which consists of a cantilevered flexible fibre, the deflection of whose end is a function of airspeed: in general the function is not linear because the drag of the fibre is not directly proportional to the velocity. The drawback of the device is the lack of any suitable recording system other than observing the end of the fibre through a microscope. A bulkier related device is the swinging-plate anemometer whose output can be read directly off a protractor scale or indirectly from an electrical circuit[39].

Vane anemometers, consisting of a windmill if the flow direction is known or the familiar array of three cups used in meteorology if it is not, can be used down to a speed limited by the friction of the bearings: these instruments are necessarily bulky.

Low rates of flow in ducts can usually be measured by means of an orifice plate with a very small hole or holes so that a measurable pressure difference is generated. If a pressure difference many times the dynamic pressure cannot be tolerated, the total *momentum* of the flow can be measured by directing the flow from the duct outlet on to a large plate placed normal to the flow, and observing the force on the plate.

Shear Stress Measurement

The shear stress in a laminar flow to which the boundary layer approximation can be applied is simply equal to the product of the velocity gradient and the molecular viscosity μ, and can therefore be obtained directly from traverses of a Pitot tube or other anemometer: the sheer stress in a turbulent flow has an additional component, the Reynolds shear stress, generated by the turbulent flow itself (see p. 28). The only direct way of measuring the Reynolds shear stress $-\rho\overline{uv}$ within the stream is with a hot wire or Doppler anemometer, but except in experiments on turbulence our interest is usually confined to the shear stress at the surface, otherwise called the surface friction (a misleading but well-established term). Because the fluctuation intensities fall to zero at the surface, the Reynolds shear stress falls to zero also, and the shear stress is equal to the product of the velocity gradient and the viscosity as in a laminar flow: the *total* shear stress remains nearly constant near the surface. Unfortunately the Reynolds shear stress is negligible only in a very thin layer, typically a fraction of a millimetre thick, adjacent to the surface and called the viscous sub-layer: it is clearly very difficult to measure the velocity gradient in this layer accurately with a Pitot tube, and even the hot wire suffers from the effect of the proximity of the wall on the heat conduction from the wire.

If we assume that the flow in a rather thicker layer next to the surface called the "inner layer" is determined entirely by the surface shear stress and the properties of the fluid, and is not influenced by the conditions further from the surface except as far as these prescribe the shear stress, we can use the method of dimen-

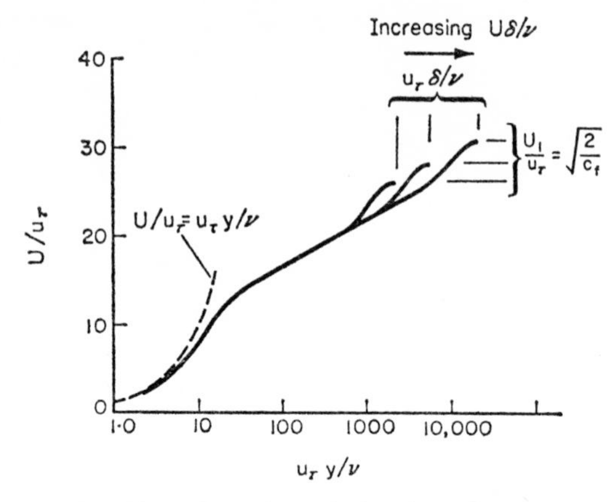

FIG. 27. "Inner law" for turbulent boundary layers.

sional analysis to show that the only velocity scale is $U_\tau = \sqrt{(\tau_w/\rho)}$ and that the only length scale is ν/U_τ. It follows that the variation of velocity with distance from the surface can be expressed as $U/U_\tau = f(U_\tau y/\nu)$ where f, according to our assumption, is a universal function, the same for any turbulent flow. In the viscous sub-layer $U = \mu y$ or $U/U_\tau = U_\tau y/\nu$ providing that $U_\tau y/\nu$ is much less than 10 (Fig. 27). This "inner law" assumption that the flow near the surface is quite independent of the outer flow cannot, of course, be exactly and perfectly true—consider the contribution to the pressure fluctuation at the surface from the eddies in the outer part of the layer or the small but finite probability that these eddies may penetrate the so-called inner or universal layer—but it seems to be a good engineering approximation for the mean velocity profile, if not for the turbulent motion itself, for $U_\tau y/\nu$ less than about 1000, the exact limit depending on the Reynolds number $U\delta/\nu$. Several methods of determining the shear stress rely on the universality of f but they depend on some initial determination of f in a flow with known shear stress: for instance it could be measured in a pipe, where the surface shear stress can be deduced from the pressure drop down the pipe, but the effect of possible departures from exact universality of f can be minimized by using an initial determination for the type of shear layer actually being investigated, such as a boundary layer. We now proceed to discuss these methods.

Surface Pitot Tubes

The Stanton tube is basically a flattened Pitot tube immersed in the viscous sub-layer of the flow: ideally, the dynamic pressure recorded by the tube should be the same function of the surface shear stress in laminar or turbulent flow. In practice it is not possible to make a tube small enough for its disturbance field to be entirely confined within the viscous sub-layer except for flows with very large values of the length scale ν/U_τ. It is therefore preferable to calibrate the tube in turbulent channel flow. The non-dimensional form of the calibration is obtained as $\Delta p/(\frac{1}{2}\rho U_\tau^2)$ or $\Delta p/\tau_w$ as a function of $U_\tau d/\nu$ where d is the height of the tube: a more convenient separation of variables is to regard $\Delta p d^2/\rho\nu^2$ as a function of $\tau_w d^2/\rho\nu^2 \equiv (U_\tau d/\nu)^2$. Instead of using a Pitot tube and a nearby static pressure tapping, any small obstacle with pressure tappings upstream and downstream can be used. The Preston tube is a larger, round surface Pitot tube, intended to occupy most of the inner layer: the advantages of greater robustness and repeatability of calibration are considerable, and the pressure differences recorded are larger, but the reliance on the universality of the function f is more complete than for the Stanton tube which does not project very far beyond the edge of the viscous sub-layer, in which the velocity profile is accurately known. Several investigators have used heat or mass transfer methods to deduce the surface shear stress: the most obvious and probably the most accurate is to use a hot wire, mounted very near the surface and calibrated *in situ* in a flow of known surface friction so that heat loss to the wall is allowed for, but measurement of heat transfer from a more robust element mounted flush with the surface is more repeatable, although in this case *most* of the heat is lost to the wall, and in either case changes in flow temperature produce errors. In water and other electrolytes, diffusion controlled chemical reactions can be used[41].

"*Floating*" *Surface Elements*

The most obvious of all methods of measuring surface shear stress is by directly measuring the force on a freely supported element of the surface. The disadvantage of the "floating" element, apart from the delicate mechanical arrangements required, is that the inevitable gap round the element is likely to disturb the flow to an unknown extent, especially if there is a longitudinal pressure gradient causing inflow or outflow through the gaps at front and rear.

Determination of Surface Friction from Overall Drag Measurements

The total drag on a body can be measured with a balance but information about the *local* surface friction at any given point on the body cannot in general be deduced from total drag measurements. One case in which it can is that of the boundary layer on a flat plate in zero pressure gradient, where the surface shear stress coefficient $c_f = \tau_w/\frac{1}{2}\rho U^2$ is a function only of the Reynolds number based on distance from the leading edge providing that the Mach number is low; if the average surface shear stress coefficient C_f is measured at different free stream speeds, c_f follows from the definition:

$$C_f = \int_0^l c_f \mathrm{d}(x/l) = \frac{\nu}{Ul} \int_0^l c_f \mathrm{d}(Ux/\nu)$$

by differentiating with respect to Ux/ν. Perhaps this is the place to remark that the flat plate is still of considerable practical interest as a test case, because such are the difficulties of surface shear stress measurement that the value of c_f as a function of Reynolds number for this, the simplest case imaginable, is still uncertain to the extent that new measurements at variance with the old by up to 10 per cent are still being published.

The total drag on a body (with the exception of any induced drag due to lift) can be found by measuring the total momentum deficit in the wake. The drag per unit span of a two-dimensional body is

$$D' = \int_{-\infty}^{\infty} [(p_\infty - p_1) + \rho_\infty U_\infty^2 - \rho_1 U_1^2] \mathrm{d}y_1 \tag{21}$$

where suffix 1 denotes the traverse station and suffix ∞ denotes conditions far upstream of the body. The integral must be taken between infinite limits and not merely across the region of reduced total pressure in the wake proper, because

$$\int_{-\infty}^{\infty} (\rho_\infty U_\infty - \rho_1 U_1) \mathrm{d}y = 0$$

by continuity, indicating that the velocity outside the wake at station 1 must exceed U_∞ to compensate for the deficit of mass flow in the wake. Several methods[8] have been developed for finding the drag from measurements within the wake only, by allowing for this mass flow deficit. The measurements of total and static pressure in the wake can be made with a "rake" of Pitot tubes and static tubes spaced across the wake, to save installing remotely controlled traverse gear for moving single probes. The surface friction drag can be obtained from the total drag by subtracting the normal-pressure drag, which is itself obtained from the integration of surface pressure measurements.

Use of the Momentum Integral Equation

The local surface shear stress coefficient can be found from measurements of the rate of increase of the momentum deficit in the boundary layer, by making traverses at two or more points at different distances downstream. The shear stress coefficient then follows from the momentum integral equation (10). Both this method and the wake traverse method suffer from the effects of secondary flow (p. 31) because any mean velocity components in the spanwise direction transfer momentum into or out of the streamwise section of the boundary layer being considered. Secondary flow is difficult to eradicate completely from a nominally two-dimensional stream, especially near side walls or in adverse pressure gradients, and for this reason it is fashionable to distrust the momentum method of measuring surface friction: the same difficulty occurs in wake traverse measurements. The most satisfying result is to obtain agreement between the momentum method and one of the other shear stress measurement methods described above: this is a reasonable guarantee that the flow and the instruments are both performing according to expectations.

Examples

1. What is the minimum speed at which a flat Pitot tube of height 0·015 cm can be used in atmospheric air if the Reynolds number is not to be less than 100?

2. Deduce from one of the equations of motion (1) that the transverse pressure gradient in a flow of radius of curvature R is given by $(1/\rho)(\partial p/\partial y) = -U\partial V/\partial x = U^2/R$, and thence find a differential equation for the static pressure in terms of the easily measured total pressure, taking the radius of curvature as known.

3. Assuming that the Nusselt number expressing the rate of heat loss from a hot wire may be approximated by a linear function of velocity over a small range, derive a differential equation connecting wire temperature and time, and verify the expression given on p. 92 for the response to a sudden small change in velocity.

4. Write down the non-dimensional parameter on which the Nusselt number chiefly depends in the case of *free* convection. (Express it as the ratio of a buoyancy force to a viscous force, and remember that there is no disposable velocity scale: assume that the fluid is a perfect gas.)

5. Suggest how very low rates of flow in ducts could be measured without uncertainties due to Reynolds number effects on the obstruction used to generate a pressure difference.

CHAPTER 4

Pressure Measurements and Manometers

Surface Pressure Measurements

Static pressure probes for use within the stream were described in the last chapter. Because the static pressure is constant across a thin shear layer of small curvature, surface pressure measurements combined with total pressure traverses are adequate for measurements of velocity in boundary layers, and surface pressure measurements can be integrated over the surface to give the overall lift, pressure drag and pitching moment on, say, an aerofoil section. Surface pressure measurement is therefore an important technique and some discussion of its difficulties may help to dispel the impression of triviality.

The measurement of pressure by tappings in the surface of a model avoids many of the difficulties associated with tubular probes in the stream, such as yaw sensitivity and the induced pressure field of the probe itself. Because the turbulent fluctuations fall to zero at the surface the only error which could be caused by fluctuations in the flow would be the result of fluctuating inflow and outflow through the hole, due to compressibility of the fluid in the leads and oscillations of the liquid level of the manometer in response to surface pressure fluctuations. The r.m.s. intensity of the surface pressure fluctuation in a turbulent boundary layer is of order $2\tau_w$ or, say, $0\cdot006.\frac{1}{2}\rho U^2$ and the effect of inflow and outflow may be expected to cancel each other to a first approximation so that errors should be small, except possibly below regions of separated flow where the pressure fluctuations may be large. The only aerodynamic difficulty remaining is the fundamental problem of the difference between the pressure on a solid surface and the pressure at the bottom of a deep hole in that surface. If we accept the inner law assumption (see p. 100) that the flow in and near the hole can be described in terms of a length scale ν/U_τ and a velocity scale U_τ we find that the pressure difference Δp can be made non-dimensional by multiplying it by $d^2/\rho\nu^2$ or dividing it by τ_w and in this form it will be a universal function of $U_\tau d/\nu$, d being the diameter of the hole.

The universal function has been determined experimentally for a turbulent flow[42], subject only to the assumption, reasonable at first sight, that the non-dimensional error is zero for an infinitely small hole. The pressure coefficient $\Delta p/(p_0 - p)$ for a two-dimensional slot in laminar flow[43] contains a term in $1/R_e$, like the drag coefficient of a sphere (p. 34) or other body, so that the error does *not* decrease to zero for zero hole size. The consequent error in the universal function for turbulent flow is believed to be small, and as the pressure differences themselves are small for reasonable sizes of hole, having been neglected altogether for about half a century without very serious results, the status of the correction is good. Typical hole sizes for low-speed experiments are 0·5 to 1 mm diameter, implying $U_\tau d/\nu < 150$ and $\Delta p/\tau_w < 0{\cdot}5$: the hole size is chosen as large as one dares in order to minimize lag in the connecting leads. In high-speed experiments the hole size is usually chosen as the smallest that can be conveniently drilled, say 0·02 mm, again implying $U_\tau d/\nu < 150$ at transonic speeds and atmospheric total pressure.

The disadvantage of static pressure tappings at fixed positions is that some interesting phenomenon, like a shock wave, invariably occurs between them. Cylindrical tubes have been used for surface pressure measurement, where the surface curvature is small and they can conveniently be traversed along the surface, but the accuracy is no better than that obtainable with tubes in the stream and there is some danger of interfering with the boundary layer.

Lag in Connecting Leads containing Gases[44]

Pressure tappings of any sort, whether on the surface of a model or in probes, must be connected to a manometer, which is usually mounted outside the tunnel or test rig necessitating the use of long pieces of connecting tubing. The tubes within the model are installed during manufacture, and are normally hypodermic tubing which can be obtained in outside diameters up to 3 mm. Flexible tubing, of translucent p.v.c. or rubber, is used for temporary external runs, but many tunnels have permanent connections from the manometers to the wall or floor of the tunnel, which can be relied on not to leak. All temporary connections must of course be leak-tested to a pressure or suction exceeding that likely to be applied during the course of the experiment: the usual method is to connect a U-tube manometer to one end of the tubing, apply a pressure difference and then seal the other end; if there is a leak the manometer reading will fall.

The disadvantage of long leads is that the pressure at the manometer will not respond immediately to changes in the pressure at the orifice, because a viscous flow through the leads is required to compensate for the change in density of the gas in the leads and the manometer which follows a change of pressure, and for the change in volume of the manometer itself if it is of the liquid level type. The former cause of lag does not arise in experiments on liquid flows. The differential equation expressing the manometer reading as a function of time and of the pressure at the orifice can be formulated by considering the mass flow through each section of the connecting tube, which is proportional to the pressure drop along the tube, and equating it to the rate of change of mass in each of the contributing volumes: if the volume of the tubes themselves is an appreciable part of the whole, as is usually the case, the calculation becomes more complicated. However, the differential equation is linear and of the first order if the pressure differences are small compared with the absolute pressure, and if we neglect the inertia of the fluids (which may be important in the case of liquid-level manometers in lightly damped systems). We can therefore define a time constant τ of the system, analogous to the time constant M of a hot wire and equal to the time which the manometer would take to adjust itself to a steep change of pressure in the orifice if the initial rate of change of manometer reading were maintained).

The time constant of the simplest system, with a tube of constant diameter d and length l leading to a U-tube manometer of cross-sectional area A and initial enclosed volume $V \gg (\pi/4)d^2l$ is

$$\tau = \left(\frac{A}{2\rho_l g} + \frac{V}{p_2}\right) \cdot \frac{128\mu l}{\pi d^4} \tag{22}$$

obtained by equating the mass flow rate through the tube, $\rho\pi d^4(p_1 - p_2)/128\mu l$, to the rate of increase of mass of gas in the manometer, $(\mathrm{d}/\mathrm{d}t)(\rho V) \equiv \rho(\mathrm{d}V/\mathrm{d}t) + (\rho V/p_2)(\mathrm{d}p_2/\mathrm{d}t)$, which is the sum of two terms expressing the rate of change of density of the gas in the manometer as a consequence of the changing pressure, and the rate of change of volume of the manometer as a consequence of the changing liquid level. Here p_1 and p_2 are the absolute pressures at the orifice and at the manometer respectively, and ρ_l is the density of the manometer liquid. The effective value of l/d^4 for a system with several sections of tube is the sum of the values for each section.

Manometers and Pressure Gauges

U-Tube Manometers

The U-tube liquid-level manometer mentioned above is simple and reliable, and apart from its contribution to connecting-lead lag the only drawback to its use is the influence of surface tension on the liquid level, which will lead to errors if the bores of the two tubes are not identical or if their walls are dirty: the two difficulties are of course connected because the bore of the manometer tube is restricted only by the need for a small volume. Alcohol is often used instead of water as a manometer liquid for gas-pressure measurement, having about one-third the surface tension.

For measuring larger pressure differences, mercury is used. The increase in specific gravity from about 1 to 13·6 leaves a range of pressures of the order of 5×10^3 to 10^4 Nm^{-2} (0·5 to 1 m water) which cannot be very conveniently measured with either liquid, but the only liquids with specific gravities of the order of $\sqrt{13}$ are corrosive and not often used. Pressures giving less than about 2 cm liquid displacement cannot be measured very accurately on a vertical U-tube manometer because the optical reading accuracy is no better than $\pm 0 \cdot 02$ cm and the uncertainty of surface tension effects makes the use of magnifying systems unwarranted unless special precautions are taken. The range can be extended downwards by a factor of 10 by using inclined tubes, but the use of angles less than about 5 deg to the horizontal again leads to surface tension troubles, even with alcohol, because the meniscus becomes extremely elongated, and in addition it is not possible to guarantee the straightness even of precision-bore glass tubing to the accuracy required. U-tubes can be used for liquid pressure measurements, either upright, with a heavy manometer liquid, or inverted, with a lighter liquid. The effective density is the difference between the manometer liquid density and the test liquid density.

Betz Manometer

A variety of liquid-level manometer which is easier to read than the basic U-tube, if no more accurate, is the Betz projection manometer (Fig. 28) in which the graduations on a scale attached to a float are projected optically on a screen. This instrument is frequently used for monitoring tunnel speed: its large internal volume makes it rather too slow in response for measurement of boundary layer profiles or sampling a series of different pressures, and its cost prohibits the use of several instruments each connected to a pressure plotting hole.

Reservoir U-Tubes

Sometimes, U-tube manometers are made with one narrow-bore and one wide-bore (reservoir) limb so that only one level has to be read, the change of level in the reservoir being negligibly small. It

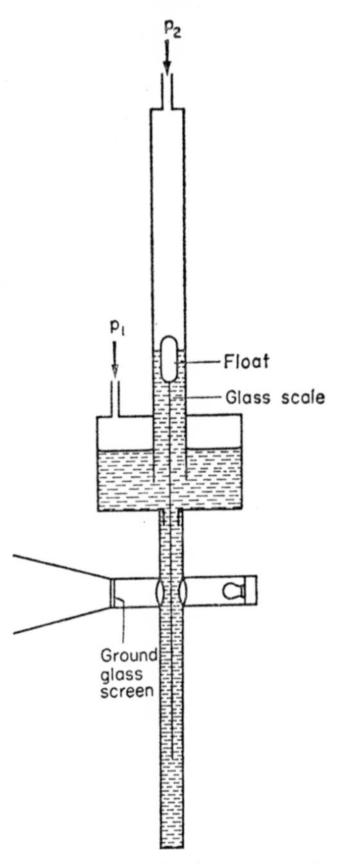

FIG. 28. Betz projection manometer.

may be necessary to make an allowance for surface tension effects, and it is not possible to make a check on the accuracy of reading the level: the sum of the readings of the two limbs of an equal-bore U-tube should be constant, providing a check on misreadings and also indicating any unevenness of bore. If the narrow-bore tube is suitably inclined and bent, it can be made to read fluid speed ($\propto \sqrt{(p_0 - p)}$) on a linear scale.

Multi-tube Manometers

Several pressure differences can be measured simultaneously with a manometer having several tubes connected at the bottom to a reservoir, also with a free surface, which is connected to a suitable reference pressure such as atmosphere, vacuum or test rig static pressure: if all the pressures to be measured differ from atmospheric by less than the range of the manometer the reservoir can simply be left open to atmosphere. For convenience the reference pressure is connected to one of the manometer tubes to avoid the difficulty of reading the reservoir level directly. The manometer tubes are clamped, side by side, to a stout base plate provided with a scale, and covered with a glass or plastic protecting plate which usually has a scale as well: the double scale helps to eliminate parallax errors. The reservoir position is adjustable and sometimes the base plate is made to tilt so that the instrument can be used inclined.

Null-displacement Manometers

Greater accuracy can be obtained with null-displacement liquid-level manometers.

The simplest arrangement of this sort (Fig. 29) is a U-tube with one limb connected at the *top* to a reservoir of adjustable height.

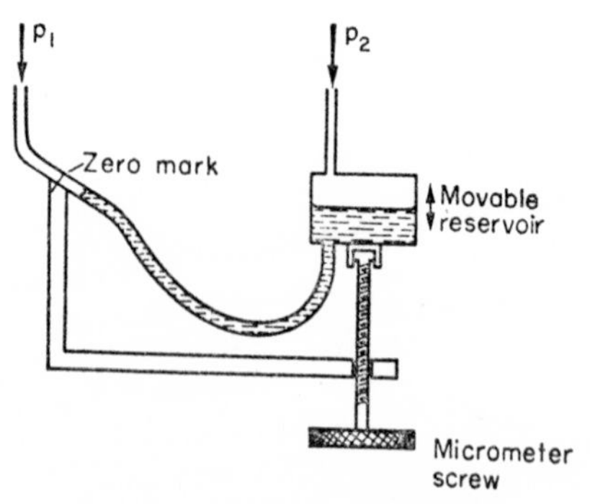

FIG. 29. Null-reading manometer.

The other limb is inclined at a small angle to the horizontal and marked at one point. The liquid level is readjusted to this point, after the application of a pressure difference, by raising or lowering the reservoir on a micrometer screw. The accuracy of the instrument is greater than that of the basic inclined-tube instrument because the meniscus position is the same for all pressure differences and so surface tension effects due to tube non-uniformity or contamination are much reduced: the use of a microscope to observe the meniscus position is justified and the accuracy may be as good as 0·0003 cm water. In the arrangement shown in Fig. 29, the connecting tube is necessarily flexible, and changes in its internal volume produce errors, but if the inclined limb is arranged in a plane normal to the paper a null reading can be obtained by tilting the entire instrument, with suitable mechanical compensation to maintain linearity of calibration. The liquid container can thus be made entirely rigid, and the instrument has even been used as a vacuum gauge: it has also been used, inverted, for water-pressure measurement. Unlike U-tube manometers, null-reading instruments can easily be converted to automatic recording by adding a shaft rotation counter or multi-turn potentiometer to the micrometer screw, although the speed of taking readings cannot be greatly increased because most of the time is needed to balance the instrument (something that a human being can do more efficiently than any automatic device). The instrument of Fig. 29 can be used as an accurate, wide-range pressure controller by presetting it to the required pressure difference and using a photocell on the zero mark to generate an error signal: the advantage over a true pressure transducer (see below) is that the same absolute accuracy can be maintained over as wide a range as necessary and that accidental overload does not cause mechanical damage.

Chattock Gauge

The Chattock gauge (Fig. 30), another null-displacement device, is basically a U-tube with a constriction between the two limbs, in which is placed a slug of some liquid immiscible with the gauging liquid: water and medicinal paraffin are the most commonly used pair. The movement of the slug is greater than the movement of the liquid–air interface by a factor equal to the ratio of the areas of the tube and the constriction, so that the smallest pressure difference that produces a detectable change in liquid level is decreased by the same factor. The usual practical form of the instrument employs a central tube in which one of the oil–water interfaces is arranged as a bubble on the end of a smaller-bore tube, and observed through a microscope. The U-tube is tilted, when a

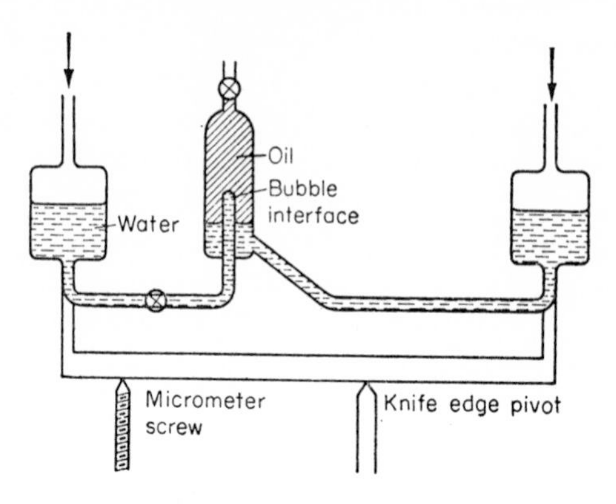

FIG. 30. Chattock gauge.

pressure difference is applied, to bring the interface back to its original position. The Chattock gauge is sensitive to contamination of the oil, to temperature differences between the two limbs, and to small changes in inclination of the table on which it is mounted: furthermore only a small out-of-balance pressure can be applied without blowing the bubble out of its tube, so that a tap is needed to close the connecting tube (Fig. 30) and must be opened with care. This instrument is used when—and only when—pressure differences must be measured to an accuracy better than 0·0002 cm water, and is now rarely found outside standards laboratories.

Pressure Transducers

These are devices for converting pressure changes into electrical signals. There are many designs, some of great ingenuity. Transducers for measuring steady pressures usually have diaphragms or bellows whose movement changes an electrical quantity, and are developments of the many types of pressure gauge in which a bellows drives a linkage and pointer. Resistance may be altered by moving the wiper of a potentiometer, or by extending a resistance strain gauge bonded to the diaphragm or stretched between it and a fixed point: reluctance or inductance can be altered by moving a slug of magnetic material or a coil: capacitance can be altered by making the diaphragm one of the plates of an air-dielectric capacitor. Similar devices can be used for fluctuating pressure or noise measurements: the moving-coil and capacitor microphones appeared in the above list. Most of the so-called mean pressure transducers will respond to fluctuations of several Hertz at least and can be used for tests on oscillating models: their other advantage over the liquid-level manometer is that, being small and remote-reading, they can be placed nearer, or actually within, the model, thus reducing connecting-lead lag. Pressure transducers are more expensive than liquid-level manometers and have to be calibrated against a standard of some sort, so that the use of one for each of a large number of pressure-plotting holes would be inconvenient: several types of scanning valve have been devised for connecting a transducer to each of a number of holes in turn while keeping the others blanked off to prevent a flow of fluid through the holes. It should be noted that if the internal volume of the transducer is small the readings can be taken at short intervals without regard for lag in the leads unless the pressures on the model change.

Pressure transducers can be used up to much higher pressures than liquid-level manometers, but their sensitivity to low pressure differences is limited by mechanical friction or thermal expansion: capacitor microphones can measure fluctuations of amplitude as low as 0·0002 cm water with ease, but could not be trusted to measure mean pressure differences of this order because of the slow drift in capacitance caused by thermal expansion. For the remote measurement of very high pressures, such as occur in the driver chambers of shock tubes, Bourdon tubes fitted with strain gauges can be used: a tube of elliptical cross-section bent into a curve tends to straighten itself out when an internal pressure is applied, and the resulting deformation can be used to extend strain gauges or to rotate a shaft or pointer as in the most familiar form of dial pressure gauge.

Measurement of Low Absolute Pressures

Very low absolute pressures, for which liquid-level manometers and ordinary barometers are insufficiently sensitive, can be measured with a MacLeod gauge (Fig. 31) which is a device for taking a sample of the gas whose pressure is to be measured and compressing it isothermally through a known volume ratio until the pressure is large enough to be recorded as a liquid level difference: the accuracy of Boyle's law is assumed. Mercury or butyl phthalate (specific gravity $\simeq 1$) are used as gauging liquids and also to provide the seal, because they have much lower vapour pressures at room temperature than any other liquids, approximately 0·001 mm Hg compared with about 10 mm Hg for water. At pressures below the vapour pressures of liquids, ionization or thermal-conductivity gauges must be used but such low pressures do not often occur even in low-density studies.

Pressure Fluctuation Measurements

Special types of pressure transducer or microphone—the terms are interchangeable though the latter is usually reserved for devices for sound measurement—are used for measurement of high-frequency fluctuations in turbulent or separated flows, and for shock tunnel measurements. There is a premium on small size, and response to steady pressures is not necessary although it facilitates calibration. Capacitor microphones are commercially available in diameters as small as $\frac{1}{4}$ in. with a frequency response extending to 100 kHz: an improved spatial resolution can be obtained by placing the microphone in a small cavity connected to the surface by a short, narrow tube (diameter as small as 0·13 mm have been used successfully) but there is bound to be an associated loss of high-frequency response, or even a resonance. It is possible to obtain a fairly flat response over an extended range of frequencies

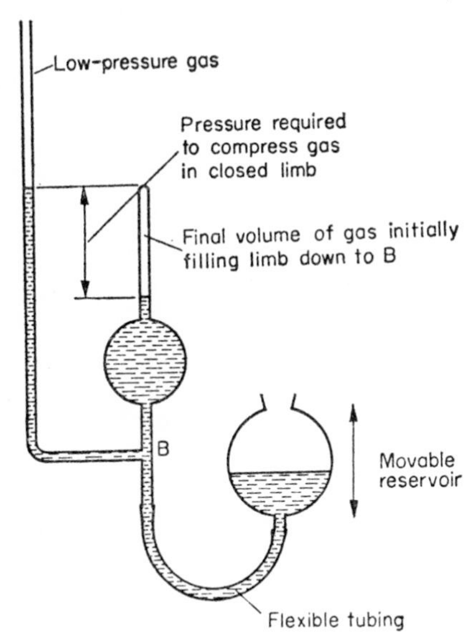

FIG. 31. MacLeod gauge.

by partially damping out the resonance, but the change of phase of the pressure fluctuation between the entry and the microphone itself may be a peculiar function of frequency. Piezoelectric ceramics have been made into pressure transducers with diameters as small as 1 mm. Quartz, and several compounds such as lead zirconate and barium titanate, generate electric charges of opposite sign at opposite faces when stressed or, in some cases, even when a hydrostatic pressure is applied: response to hydrostatic pressure is not necessary as it is usual to cover the sensitive crystal with a diaphragm flush with the surface.

Examples

1. A Pitot tube $2 \cdot 5$ cm long and $0 \cdot 037$ cm in internal diameter is connected by $1 \cdot 5$ m of 5 mm bore tubing to a water U-tube manometer whose tubes are of 8 mm bore, the liquid level being initially 25 cm below the top of the tube. After verifying that the value of l/d^4 for the connecting tube is very much smaller than that for the Pitot tube so that the former contributes to the volume of the system but not noticeably to the pressure drop, calculate the time constant of the system in response to a small change in pressure at the mouth of the Pitot tube. (Take the viscosity of air as $1 \cdot 78 \times 10^{-5}$ kg m^{-1} s^{-1}.)

2. Calculate the sensitivity of the manometer of Fig. 29 in terms of change in meniscus position per unit pressure difference.

3. If the correction to surface pressure readings for finite hole size is so small (of the order of $0 \cdot 002 \times \frac{1}{2}\rho U^2$) why is there any need to keep the hole diameter small?

CHAPTER 5

Force and Position Measurements

THE derivation of the resultant of the normal forces on a body from the integration of measured pressure distributions was mentioned in the last chapter. Forces can be directly measured by either of two types of mechanism, called the true (or null-displacement) balance and the strain balance, and exemplified by the chemical balance and the spring balance respectively. In both cases it is necessary to allow the body experiencing the force to move, perhaps infinitesimally, in the direction of application of the force. Balance measurements are routine in aeronautics and ship-tank testing, whereas in most other branches of the subject they are made only occasionally for special purposes. The discussion below refers chiefly to wind tunnel balances but, of course, the general principles are of wider application.

The Null-displacement Method

In this method, an opposing force is applied, typically by means of weights or an electromagnet, so as to bring the body back to its original position of equilibrium before the application of the forces. The sensitivity, or least detectable reading, of the balance depends on the force required to disturb it noticeably from the original position of equilibrium, and can be defined as $\delta x/(\mathrm{d}x/\mathrm{d}F)$ where δx is the least displacement that can be detected and $\mathrm{d}x/\mathrm{d}F$ is the displacement per unit applied force. This can be seen qualitatively from a consideration of the chemical balance. It can also be seen that balances for experiments on fluid mechanics must not have too large a value of $\mathrm{d}x/\mathrm{d}F$ or the body may move sufficiently for the forces or moments on it to alter appreciably. Some form of damping is always needed for a null-displacement balance to help reduce the effects of unsteadiness in the flow round the body or in the test-rig stream itself.

The Strain Method

In this case, the displacement of the body or some part of the balance structure in response to an applied force is measured directly, and the sensitivity can again be defined as $\delta x/(\mathrm{d}x/\mathrm{d}F)$. If the force is to be measured to an accuracy of say 1 per cent, the displacement under the load must be at least $100\delta x$ since δx is defined as the least displacement that can be detected: since the maximum tolerable displacement depends only on the model and not on the details of the force-measuring method, the accuracy of displacement measurement must, other things being equal, be something like a hundred times better in the strain balance than the indication of equilibrium position in the null-displacement balance.

Either method could be used with almost any balance framework and linkage, and it is largely an historical accident that the two methods as applied to fluid mechanics are usually associated respectively with the permanent balances of low-speed tunnels and the sting balances (p. 127) of high-speed tunnels. Most special-purpose balances use the strain method and it is likely to supersede the null-displacement balance for most purposes in the future.

Forces on Tunnel Models

The force on a model can be divided into components along three mutually perpendicular axes through any chosen point, together with three moments about the same axes. It is very seldom that all six components have to be measured in the same experiment and only the larger wind tunnels have permanent six-component balances. In a permanent balance one axis is chosen to be in the direction of the tunnel stream and another is chosen to be vertical so that the weight of the model has to be subtracted from, or counter-balanced in, only one component of force and affects only one moment: sting balances necessarily pitch, yaw and roll with the model and one of the axes is chosen to coincide with the centre line of the model.

These two systems of axes are called wind axes and body axes respectively, but it is the usual practice to convert all measured

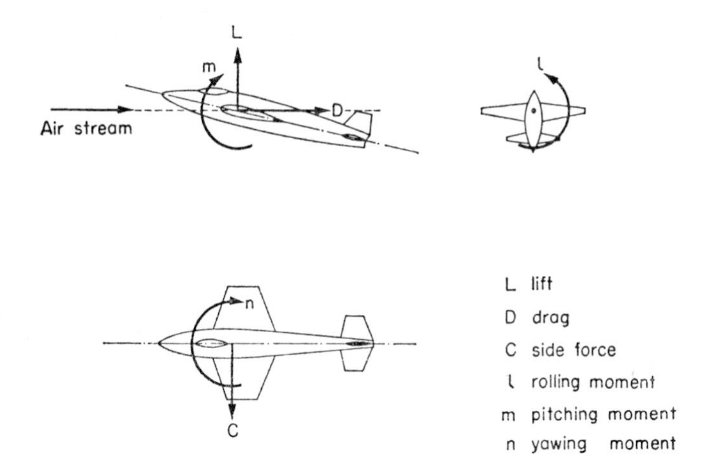

FIG. 32. System of axes for force and moment measurement.

forces and moments to wind axes, when they are called lift, drag, and side-force, and yawing, rolling, and pitching moment (Fig. 32). In steady flight the last four are zero and the first two are balanced by the weight and thrust, but in wind tunnel experiments on complete aircraft models it is usually more convenient to measure the forces caused by a change of attitude directly and to determine the control surface movements required to produce trimming forces later, rather than to adjust the control surfaces to trim the model while the tunnel is running.

Permanent Wind Tunnel Balances[8]

The fundamental difficulty in the design of permanent balances is that they may have to measure forces and moments to acceptable *percentage* accuracy in the full range from the maximum capacity of the balance (defined as the largest load that can be applied to the balance before unacceptable distortions of the balance structure occur) down to a very small fraction of this. For instance an experiment on a laminar-flow suction aerofoil might be followed by tests on a ground-installation radar aerial or some other extremely bluff body, and the same drag balance might have to cater for both. This difficulty arises with most measuring instruments, such as manometers which may be required to measure either the tunnel dynamic pressure or the pressure difference produced by a Stanton surface tube, perhaps a few thousandths of the dynamic pressure: however, most laboratories have several different types of manometer available for different pressure ranges but few have interchangeable null-displacement balances. Although the actual weighing units or "weighbeams" of permanent balances could perhaps be changed from experiment to experiment according to the capacity required, the mechanical linkages of the balance would still have to move freely under the smallest load and not deflect noticeably under the largest, so that little would be gained. Permanent balances are therefore designed to have a sensitivity of as little as 10^{-4} of the maximum capacity.

Resolution of Forces

The six components are not invariably measured as six separate corresponding forces on the balance: the simplest type of lift and pitching moment balance has two measuring weighbeams, fore and aft (Fig. 33), the lift being obtained as the sum of the readings and the pitching moment as a quantity proportional to their difference. Furthermore, the balance design may be such that only one weighbeam can be operated at one time, the others being locked. Interactions may occur between allegedly perpendicular components so that, for instance, the application of a pure drag force may affect the lift reading slightly: the best that can be hoped is that these interactions should be small enough to be assumed linear, so that the *percentage* of the drag force appearing in the lift reading shall be independent of the lift and drag.

The greatest simplicity of balance operation generally involves the greatest degree of mechanical complication and probably also a sacrifice of accuracy. The modern trend is towards simultaneous measurement of all the components by automatic weighbeams so

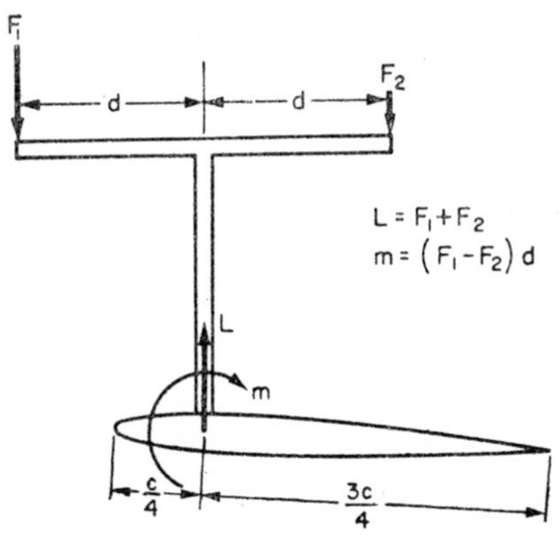

FIG. 33. Lift and pitching moment balance.

that automatic data recording techniques can be used. If the results are to be reduced to coefficient form by an electronic computer it is a small matter to add two weighbeam readings to obtain one component, or to allow for interactions between the components, although the calibration of the balance to discover the interactions may still be tedious.

Balance Linkages

Practically no two balances employ the same linkage system and a full discussion of all the ingenious mechanisms employed in the past would occupy much space. Most of the linkages can be classified into two types, the virtual-centre balance and the parallel-linkage balance. In the virtual-centre balance, moments are measured by allowing the model freedom to rotate about the appropriate axis by attaching it to a trapezoidal linkage, the pro-

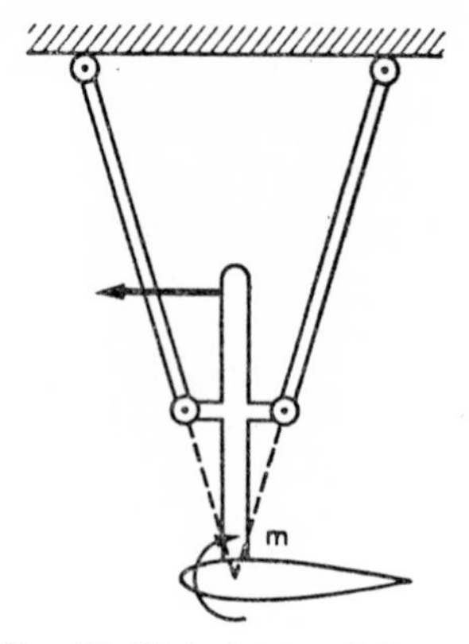

FIG. 34. Virtual-centre balance.

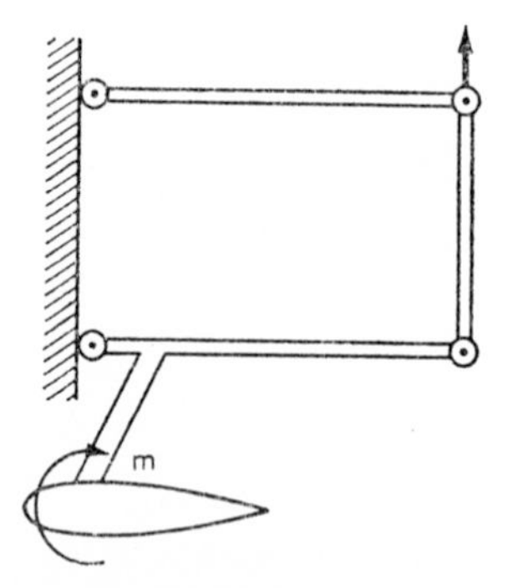

FIG. 35. Parallel-linkage balance.

jections of whose inclined members meet on the axis (Fig. 34). The moment can then be obtained from the reading of a single weigh-beam. In the parallel-linkage balance the "virtual centre" is at infinity (Fig. 35), the linkage is rectangular instead of trapezoidal, and the moment is obtained in terms of the force on the vertical links: the moment can be measured with one weighbeam at the expense of needing two for the force balance unless a further stage of linkage is employed. A slightly different principle is that of the multi-moment balance, in which lift, drag and pitching moment are measured in terms of moments about three axes, the model and its supports being allowed freedom to rotate about each axis in turn, giving three simultaneous equations for the three components.

Weighbeams

Weighbeams are usually attached to the balance by bellcranks or other suitable linkages so that the displacement of the balance indicator is increased by a large factor over the displacement of the main part of the balance structure, and the force which must be applied to the weighbeam correspondingly reduced from the actual force exerted on the model. This makes operation of the balance easier, particularly if the weighbeam is of the simple steelyard type (Fig. 36(a)) on which weights are hung or slid by the operator, but makes increased demands on the rigidity of the linkage. The displacement can be indicated by an optical projection system or by an electrical signal derived by any of the strain gauge principles mentioned on p. 115, generally a variable inductance. The steelyard is not particularly well suited to automatic balancing, although it has the advantage of relying on the invariability of weights rather than any more changeable quantity. Self-balancing steelyards have been used at the R.A.E. for many years, using a leadscrew-driven sliding weight and a cam mechanism to add or remove dropweights at either end of the range of travel of the sliding weight.

Kelvin current balances (Fig. 36(b)), in which the balancing force is applied by coils carrying an electric current, are more suitable for automatic recording. Two fixed coils, carrying the same current, produce a magnetic field which is roughly constant along the length of the gap between them, and a third coil, attached to the weighbeam and also carrying a current, is inserted in the gap. A particular advantage of this arrangement is that the currents in the fixed and moving coils can be varied independently over the full range in which they can be measured accurately, so that the ratio of full-scale reading to sensitivity of the weighbeam is the square of the ratio of full-scale reading to sensitivity of the ammeters: the current in one coil can be varied in pre-calibrated steps to give different ranges of sensitivity characterized by

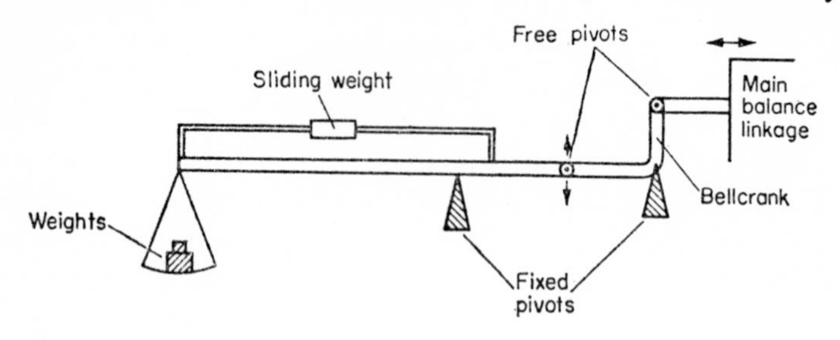

(a) Steelyard weighbeam

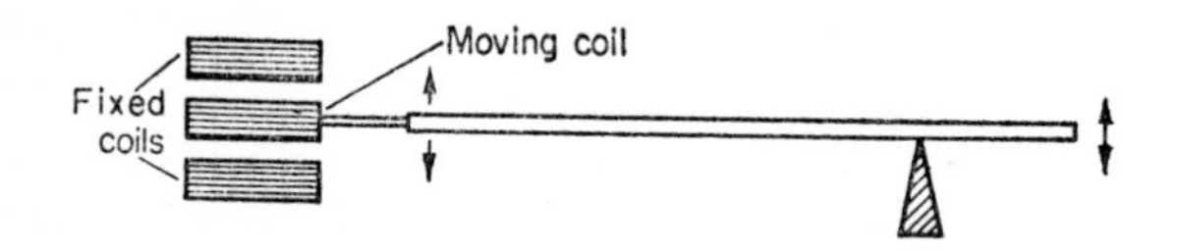

(b) Current–balance weighbeam

FIG. 36. Balance weighbeams.

different calibration factors. However, the accuracy and reliability of current balances is in practice somewhat inferior to that of the best steelyard types.

Other types of weighbeam include the pressure capsule or load cell, in which the force is balanced by a fluid pressure read on a manometer but which is likely to be somewhat inferior in range to the other types mentioned, and the strain gauge which suffers from the same disadvantage and is more frequently used in the strain type of balance.

Model Supports

It is usual for the balance linkage to be mounted entirely outside the tunnel, the attachments to the model being made by wires under tension or by rigid members pin-jointed at their ends. The minimum possible number of supports, namely three, is used, one on either side of the centre line on or near the lateral axis, and one near the rear of the model which is lengthened or shortened to change the incidence. Wire supports have the minimum drag but are obviously unsuitable for supporting a model from below: although wires are less popular than in the earlier days of wind tunnel testing, many existing tunnel balances are mounted on top of the tunnel. In order that the tension in the wires shall not fall to zero, the model is usually mounted with the lift force acting downwards.

Strain Balances[11]

Strain gauges can be used with a linkage like that of a null-displacement balance but in aeronautics are more commonly used in sting balances, which are effectively rigid members, attached to the rear of the model and aligned approximately in the stream direction to reduce interference. The rear end of the sting is attached to a beam spanning the tunnel which is tilted to change the incidence of the model. The sting support method is preferred to wires or rigid supports for tests in high-speed flow because of the large size and uncertainty of interference corrections in transonic and supersonic flow: in low-speed flows it is usually simple enough to make a correction for the drag of the struts and their effect on the flow round the model. The interference caused by a sting support is usually neglected, justifiably unless the sting is large enough to cause noticeable changes in the pressure on the rear of the model through upstream influence in the subsonic part of the wake. Stings have occasionally been used to mount high-speed tunnel models on null-displacement balances, as in the Caltech. Jet Propulsion Laboratory 18 $\times$ 20 in. supersonic tunnel which has a six-component virtual-centre balance.

Resistance Strain Gauges

The gauges are invariably the resistance type, in which an elongation caused by a strain in the member on which the gauge is mounted produces a change in electrical resistance of the gauge. The percentage change in resistance is a multiple of the percentage strain called the gauge factor, which is normally about two for the wire type of strain gauge. The strain measured can be related to the force or bending moment on the rigid member. The wire is usually doubled back and forth to increase the effective length and attached to a thin sheet of bakelite material, and the assembly cemented to the rigid member. Gauge lengths of 0·5 to 1 cm are typical: much smaller gauges are used in structural analysis. Semi-conducting materials have a much higher gauge factor than metal, but are more common in pressure transducers than in force balances.

Resolution of Forces

Lift and pitching moment can be measured on a sting balance in the same way as in the multi-moment balance mentioned above by measuring the bending moments, M_1 and M_2 say, at two points distances x_1 and x_2 from the origin of the force axes. The lift, or strictly the force normal to the line joining the two points, is then $(M_1 - M_2)/(x_1 - x_2)$: pitching moment is $(M_2x_1 - M_1x_2)/(x_1 - x_2)$. Side force and yawing moment can be measured similarly. The bending moment at a point on the sting is proportional to the difference of the readings of two strain gauges, assumed to be identical, on the tension and compression sides of the sting beam, which may conveniently be a hollow tube as shown in Fig. 37. The drag can be obtained as the sum of the readings of two such gauges, but a little thought will show that the strain produced by the drag will generally be much smaller than the strain produced by the lift force, and in practice it is usually expedient to use entirely separate gauges for drag measurement. The arrangement shown in Fig. 37 has gauges on the flexures of a parallel-flexure linkage so that an adequately large strain can be produced by the drag force without large displacements being produced by the lift force: the flexures are merely extended or compressed by the lift and pitching moment, but the drag applies a bending moment to them.

Rolling moment can be measured by one or more pairs of gauges arranged at plus and minus 45 deg to the sting axis: the torsional moment is proportional to the difference between the strains in these two principal directions. Alternatively, rolling moment can be measured in terms of the extension and compression of links either side of the centre line of the sting. However, the

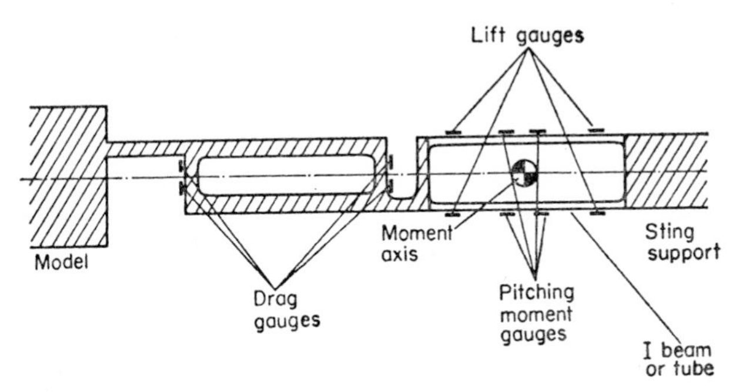

FIG. 37. Strain gauge sting balance.

usual procedure is to attach the rear of the sting to the tunnel via a journal bearing, allowing it freedom in roll, and restraining it from rolling by a member to which strain gauges are attached.

Practical balances rarely have gauges for all six components because they are normally designed for a particular model. The gauge arrangements may be more complicated than those outlined above, partly because of the need to compensate for the temperature sensitivity of the gauges. In addition the gauges cannot be assumed to be identical though quality control of a given batch is good, and careful calibration of the balance is necessary: quite complicated arrangements may be needed to calibrate a six-component balance.

Temperature Sensitivity

A temperature change of 1°C produces the same change in resistance as a strain numerically equal to α/G where α is the temperature coefficient of resistance and G is the gauge factor. If $\alpha = 1/300$, the approximate figure for most pure metals, the apparent strain would be of the order of 1/300, implying a stress of the order of 10^5 lb/in^2 in steel, considerably more than could safely be applied to a sting: this magnitude of error can be greatly reduced by using a suitable alloy material for the wire. Maximum strains are about 0·0005 to 0·001 (implying a stress of about 10^8 to 2×10^8 N m^{-2} in steel), so that a low temperature coefficient of resistance must take precedence over other desirable qualities for a strain gauge material, being the main reason for the lack of popularity of semi-conductor gauges, but further precautions are necessary to keep errors due to changes in temperature to a minimum. The simplest method of temperature compensation is to suspend a second, unstrained, gauge near every "active" gauge and to connect the two gauges in adjacent sides of a Wheatstone bridge so that temperature changes do not alter the balance of the bridge. If the outputs of two adjacent active gauges are to be subtracted to obtain a force or moment, changes in resistance due to temperature will cancel out. Present-day practice is to use four active gauges in a bridge so that temperature differences between separated pairs of gauges cancel out, and spurious strains can be reduced to the order of $0 \cdot 15 \times 10^{-6}$/°C, or 0·15 "microstrain" /°C. Balances for intermittent-running tunnels and shock tubes can be thermally insulated (balances have always to be enclosed within guards so that no air loads act upon them) and air-conditioned balances have been used in continuous-running tunnels, particularly those which experience large stagnation temperature changes, but these precautions are to be regarded as additional to the use of a balanced bridge of gauges using a material of low temperature coefficient of resistance.

Strain gauges are also sensitive to humidity in a rather erratic fashion. An argument against trying to waterproof the individual gauges is that they are then harder to dry out when they *do* get damp, but if trouble is anticipated it is possible to seal up the complete balance, with a desiccant, in its guard. The problem is obviously more severe in balances for use in or near water: strain gauge balances are now widely used in ship-tank and propeller testing.

Frequency Response

Strain gauge balances have the advantage over null-displacement balances of a very much faster response: even if the weigh-beam of a null-displacement balance is quick-acting, the natural frequencies of oscillation of the balance about the null position are likely to be far too long for rapidly changing forces to be measured. The only limit to the frequency of response of a strain gauge balance is set by the natural frequencies of the sting and model (see p. 73). As well as being used for the measurement of oscillating forces on complete models, strain gauges can be used for measuring control hinge moments by attaching them directly to the hinges; before the introduction of resistance strain gauges, steady hinge moments were measured by connecting the control surface to a balance outside the tunnel and unsteady hinge moments could not be measured by any convenient method.

In order to minimize the effect of drift of d.c. output amplifiers, strain gauge bridges are sometimes operated on alternating current at a frequency which must exceed, by a factor of the order of ten, the maximum frequency of response required so that the excitation voltage frequency can be filtered from the output without affecting the output proper. The great strides made in d.c. amplifier design over the last few years have resulted in a general return to d.c. excitation, which is obviously simpler in principle.

Tunnel Interference[45]

The theory of the corrections to be applied to force or pressure measurements in wind or water tunnels because of the finite extent of the stream is complicated, inaccurate and—to most people—uninteresting. Unfortunately it is often essential to make these corrections, and every tunnel user must therefore understand them sufficiently well to apply them. The explanation which follows will, to avoid confusion, be restricted to incompressible flow. Any given correction for a Mach number below choking, with the exception of the blockage correction to the density, can be obtained from the corresponding correction for incompressible flow by multiplying by some power of $\sqrt{(1 - M^2)}$, a factor which constantly occurs in discussions of subsonic compressible flow and is denoted by the symbol β. In supersonic flow all the corrections will be negligible if the reflection of the model's bow shock wave from the tunnel wall does not hit the model or greatly disturb the wake. At transonic speeds, where the bow shock would hit the model again even if choking of the tunnel did not occur, perforated or slotted walls are used to minimize the strength of the reflection, and the calculation of the interference corrections becomes very complicated indeed.

The corrections for incompressible flow are nearly all approximate, and they are assumed not to interact with one another: this means that the corrections must be kept small by making the model reasonably small compared with the size of the tunnel. If the corrections are kept down to about 10 per cent of the uncorrected reading and are themselves known to an accuracy of 10 per cent the accuracy of the final result will be 1 per cent, which is good enough for most purposes, and does not encourage effort to improve the accuracy of the corrections: however, if the corrections were any larger or less accurately known than this one would start to feel unhappy about the accuracy of the results. The problem has become more acute with short or vertical take-off STOL or VTOL aircraft which, at some parts of their transition to horizontal flight, may operate at very high lift coefficients and produce large deflections of the flow: some progress has been made with the extension of classical corrections to cases such as the jet flap (another high-lift case), the jet sheet being replaced by an inviscid vortex sheet, but viscous effects may make this sort of approximation very doubtful, particularly in cases where the model is deliberately mounted near the tunnel wall to simulate the effects of proximity to the ground.

Solid Blockage

The most obvious effect of tunnel wall constraint on the flow around a body is to increase the average tunnel velocity, from U_∞ to $(1 + \epsilon)U_\infty$ say, to compensate for the reduction in effective

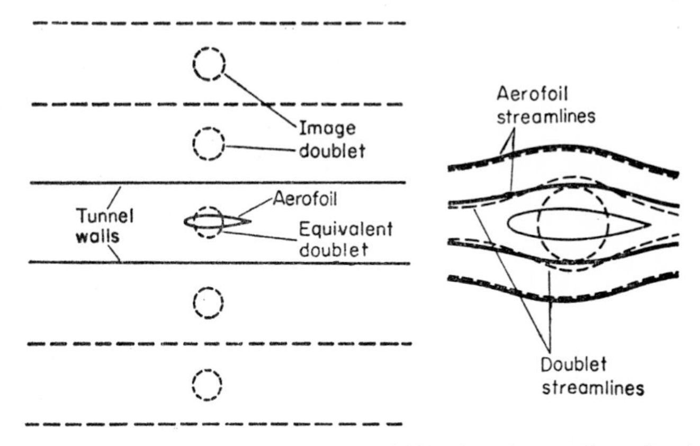

FIG. 38. Image system representing solid blockage in two-dimensional flow.

cross-sectional area of the tunnel by the blockage caused by the model. Because the length of the model, as well as its thickness, is small compared with the width of the tunnel, a one-dimensional theory will not suffice. The model is replaced by a doublet producing the same flow field at large distances, and the effect of the tunnel walls is simulated by an infinite array of "image" doublets: in two-dimensional flow an infinite number of parallel line doublets is used and the tunnel walls correspond to the lines of symmetry between adjacent doublets (Fig. 38). In three-dimensional flow the arrangement of image doublets, and the resulting algebra, may become complicated.

In the case of a symmetrical two-dimensional aerofoil in a rectangular tunnel the velocity near the model is increased by a

$$\epsilon = \frac{\pi}{3h^2} \int \frac{U(s)}{U_\infty} y \, ds \tag{23}$$

fraction where h is the height of the tunnel, y the aerofoil ordinate, and $U(s)$ the velocity near the aerofoil surface, the integral being evaluated over one-half of the aerofoil profile, and the arc s being measured over the surface of the aerofoil: for thin aerofoils, s can be measured with sufficient accuracy along the chord line. Tabulated values of the integral are available for various aerofoil sections, and if we consider only thin aerofoils (thickness/chord ratio less than 0·1, say) the integral can be evaluated from the linearized theory of thin aerofoils, in which case the fractional increase in velocity is

$$\epsilon = \frac{\pi A}{\mathrm{b}h^2}\left(k_0 + k_1 \frac{c}{t}\right) \text{ or } k\left(\frac{c}{h}\right)^2\left(\frac{t}{c}\right)\left(1 + \frac{k_1}{k_0}\left(\frac{t}{c}\right)\right) \tag{24}$$

where A is the area of the aerofoil *section* and t its maximum thickness and k_0 and k_1 are constants of the order of unity, the exact values of k_0, k_1 and k depending on the type of profile.[8] When the aerofoil is at incidence this solid blockage is increased by $(c/h)^2\alpha^2 f(t/c)$ where α is the incidence and f is a universal function of t/c. Models consisting of combinations of aerofoils and bodies of revolution can be treated by superposing the blockages caused by the various components.

Wake Blockage

Because of the momentum deficit in the wake, which is proportional to the drag of the model, the velocity in the flow outside the wake downstream of the model must increase to maintain constant mass flow: to a first order of approximation, based on the representation of the wake by a source at the model, it can be taken that half of this velocity increase has already occurred upstream of the model. In two-dimensional flow, the momentum deficit (see eqn. (21)) is given by

$$D' \equiv C_D \tfrac{1}{2} \rho U_\infty^2 c$$

$$= \int_{-\infty}^{\infty} \rho U(U_\infty - U)\, \mathrm{d}y \simeq \rho U_\infty \int_{-\infty}^{\infty} (U_\infty - U)\, \mathrm{d}y \quad (25)$$

(the approximation being quite good for a typical aerofoil because the velocity in the wake differs from the free stream velocity by only a few per cent at more than one or two chord lengths downstream). The increase in the external flow velocity over the rest of the tunnel height needed to compensate for the mass flow deficit is therefore independent of downstream distance (to a one-dimensional approximation) and given by $h\epsilon U_\infty = \int_{-\infty}^{\infty}(U_\infty - U)\mathrm{d}y$ so that half this increase, the quantity to be added to the measured velocity far upstream, is equal to $C_D c U_\infty/4h$. There is also an induced velocity *gradient* of the order of $C_D U_\infty c/h^2$ which is negligibly small unless C_D is large or the model is too large for the tunnel. For bluff bodies[46], where nearly all the drag is base drag, the factor 1/4 is nearer 1/2 for two-dimensional flow and 5/4 for aspect ratios up to 10.

Tunnel Wall Boundary-layer Blockage

The growth of the tunnel wall boundary layers causes an increase in velocity down the working section unless the tunnel walls or corner fillets taper to allow for this growth. The resulting longitudinal pressure gradient causes a (horizontal) "buoyancy" force on the model which, like the buoyancy force on an object immersed in a liquid, is equal to the volume of the body times the pressure gradient: in two-dimensional flow, the drag per unit span depends on the cross-sectional area of the body instead of the volume. The longitudinal buoyancy correction, to be added to the measured drag coefficient, is $(-\mathrm{d}C_p/\mathrm{d}x)(A/c)$ where A is the volume or cross-sectional area of the model, C_p is the usual pressure coefficient $(p - p_{ref})/(p_0 - p)$, and c is the area or length on which C_D is based.

The other corrections mentioned above have also to be applied to three-dimensional models, but the formulae are usually more complicated and will not be discussed in this short account.

Lift Interference (Fig. 39)

The effect of tunnel wall constraint on a lifting model can be simulated in two-dimensional flow by replacing the model by a

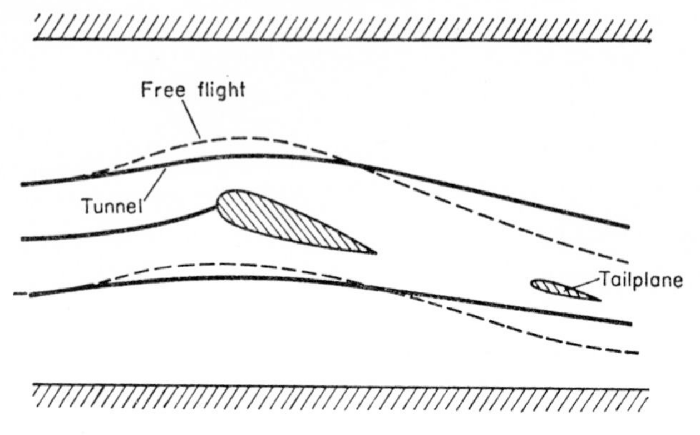

FIG. 39. Effect of tunnel constraint in reducing streamline deflection (incidence) and curvature (camber).

spanwise line vortex at the centre of lift and representing the wall constraint by images as in the case of solid blockage. It is found that there is a change of effective incidence which depends on the pitching moment as well as on the lift, because the vortex must be placed at the centre of lift, which does not coincide with the origin of the axes unless the pitching moment happens to be zero. There is also a change of effective camber which depends only on the lift, that is to say on the circulation round the aerofoil which is equal to the strength of the image vortices. The usual practice is to use the theory for a thin circular-arc aerofoil to find the changes in lift and pitching moment produced by the change in camber angle, and to apply the corrections to the measured C_L and C_m. The incidence correction can also be converted into corrections for C_L, C_m and C_D, but this is not usually done in practice. The reason is that inviscid aerofoil theory, which would have to be used to evaluate the gradients of the force coefficients with incidence, would be highly dubious near the stall, where $\partial C_L/\partial \alpha$, for instance, is very much less than the inviscid-theory value of 2π per radian. Of course, the theoretical corrections for induced camber become inaccurate near the stall as well, but it is better to make empirical allowances for this than to end up with results applying to an aerofoil of different camber from that actually tested.

In three-dimensional flow round a lifting body, the trailing vortices (see Plate 2), as well as the bound vortex line representing the wing itself, must be included in the image system. If the span of the wing is not small compared with the width of the tunnel, the induced camber and incidence will vary across the span: these induced quantities are in addition to the usual finite-wing effects which occur in an unrestricted stream. It is not possible to make any sense out of corrections which vary across the span without making rather crude assumptions about the variation of loading over the span of the wing, and the equally crude assumption that the corrections can be averaged over the span of the wing is much easier to apply. As the chord of three-dimensional test wings is usually much less than that of two-dimensional aerofoils, the corrections due to the images of the bound vortex, which are of order $(c/h)^2$, can often be neglected, leaving only the trailing vortices to be taken into account.

The numerical values of the lift corrections to camber angle γ and incidence α in two-dimensional flow are $\delta\gamma = (\pi/192)(c/h)^2 C_L$ and $\delta\alpha = (\pi/96)(c/h)^2(C_L + 4C_m)$: the rates of change of C_L and C_m with γ are respectively 4π and $-\pi$ per radian. As an example, the incidence correction for an aerofoil with $c/h = \frac{1}{2}, C_L = 1$ and $C_m = 0$ is $\delta\alpha = \frac{1}{2}$ deg: the camber corrections to C_L and C_m are respectively $0{\cdot}05$ and $-0{\cdot}012$.

A final correction due to tunnel constraint on lift is to the downwash angle over the tailplane, which is most easily applied as a correction to the measured pitching moment unless the tailplane setting is actually to be varied during the course of the experiment: the correction is derived from the same image system as that used to calculate the incidence correction.

Aspect Ratio Corrections

It is sometimes necessary to correct the results of tests on wings of finite aspect ratio, A say, to obtain data for the same aerofoil section in two-dimensional flow, and the aircraft designer has frequently to make the reverse calculation. In the early days of tunnel testing, most measurements of section characteristics were made with rectangular wings of aspect ratio 6, but more recent tests have usually been made with wings spanning the tunnel, often with only the central portion of the wing mounted on the balance and aligned with dummy sections attached to the tunnel wall, to minimize the effect of the tunnel wall boundary layers. Corrections to results of tests on finite wings are obtained by considering the induced velocity field of the trailing vortices and therefore depend on the spanwise loading (that is, on the planform if the wing is untwisted and of constant section). Corrections are usually made to C_D, α and $\partial C_L/\partial\alpha$ at constant C_L and are

$$\delta C_D = \frac{-C_L^2}{\pi A}(1+\sigma) \tag{26}$$

$$\delta\alpha = \frac{-C_L}{\pi A}(1+\tau) \tag{27}$$

$$\delta[1/(\partial C_L/(\partial\alpha)] = \frac{-1}{\pi A}(1+\tau) \tag{28}$$

where σ and τ are zero for elliptical wings and are in general functions of planform and $A/(\partial C_L/\partial\alpha)$. For a rectangular wing of aspect ratio 6 and $(\partial C_L/\partial\alpha)_{A=\infty} = 2\pi$, we have $\sigma = 0{\cdot}03$ and $\tau = 0{\cdot}12$ so that $\partial C_L/\partial\alpha = 2\pi/1{\cdot}37$: it will be seen that the lift curve slope is a good deal below the two-dimensional value even for what is nowadays a moderately high aspect ratio.

Corrections to Surface Pressure Distributions

The most frequent application of corrections for tunnel constraint is to the overall forces and moments, but it is clear that the surface pressure distributions, which can be integrated to give forces and moments, should also be corrected. The allowances for blockage are again regarded as changes in effective tunnel speed, the correction to incidence is again applied as such, and the only correction that presents any new difficulty is that for the induced camber. If we invoke the principle of superposition, we see that the "supervelocity" $U - U_\infty$ at any point on the aerofoil will be equal to the sum of the supervelocity on the same aerofoil in unrestricted flow and the supervelocity on a circular-arc aerofoil, with camber equal to the induced camber, at *zero* incidence. Denoting the latter by δU (a function of x/c) we find that the correction to be added to the measured pressure coefficient $C_p \equiv 1 - (U/U_\infty)^2$ at any point is $2\sqrt{(1 - C_p)}\,(\delta U/U_\infty)$.

Permissible Model Size for a Given Tunnel

There is no short answer to the question of permissible model size. Formally one should estimate the various corrections using values of the force coefficients obtained in previous tests on similar models, decide how large a correction can be applied before the accuracy of the corrected result falls below that desired, and thence deduce the maximum permissible ratio of span or chord to tunnel size. Unless it is desired to use the maximum possible size of model, the usual practice is to choose a model size which seems likely, again on the basis of previous experience, to require only reasonably small corrections. The ratio of model size to tunnel size actually used varies considerably with the different interpretations of the phrase "reasonably small". It must be remembered that the usual reason for wanting a large model is to obtain the maximum possible Reynolds number, or occasionally to simplify the construction of the model: thus, a 20 per cent difference in model size is not likely to mean the difference between success and failure. Occasionally the model size may be fixed beforehand, if, for instance, tests of full-scale objects are being made, and it may then be necessary to make calculations to decide if one's tunnel is large enough to take the model.

Position Measurements

The strain gauge is nominally a position indicator and is used as such in the diaphragm-type pressure transducer. Other position measurements, such as measurement of model attitude or probe position, are made by conventional methods, not peculiar to fluid mechanics. An ever-recurring problem with no universal solution is the support and positioning of Pitot tubes and hot wires for shear layer exploration. The position of the probe must be controllable to within one or two thousandths of a centimetre, the probe must not vibrate (especially in the case of hot wires for fluctuation measurements) and the support structure must be slender enough not to interfere with the flow: a particular danger is that boundary-layer separation may be provoked or worsened. Control is usually obtained by means of a micrometer-type screw thread driven by a flexible cable: the preferable alternative of a rigid drive shaft is only usable if a limited number of traverse stations is sufficient. It is best of all, from the point of view of accuracy, to mount the drive motor on the end of the screw thread, but the bulk of the motor and remote-reading position indicator is usually unacceptable if the whole device must be placed in the stream. Relative vibration is minimized if the probe is attached to the model, but if it is felt to be essential to avoid any blockage in the boundary layer a gantry attached to the tunnel must be used. In this latter case, maximum strength with minimum blockage is achieved by using a support which decreases in diameter from its root to its tip where the probe is mounted.

The deduction of forces from measurements of acceleration has been mentioned briefly in Chapter 2, p. 37: this method would not be used if simpler alternatives were available, and no novel techniques of position or acceleration measurement have so far been developed for shock tunnel use. The forces on free-flight models, particularly the drag of projectiles, are commonly obtained from accelerations measured by theodolite cine cameras, radar or the doppler effect on the frequency of a radio transmitter in the model as observed from the ground.

Surface position measurements are needed in studies of surface waves in liquids. For long wavelengths or tidal measurements, a float attached to a potentiometer or other electrical recording device can be used: for shorter wavelengths (higher frequencies) an electrical circuit containing a resistance or capacitance modified by the height of the liquid is more suitable. The simplest such device is a vertical resistance wire passing through the water surface, the circuit being completed by the water itself so that the resistance varies linearly with the length of wire above the surface. Alternatively the capacitance a vertical electrode (insulated from the liquid) can be measured in an a.c. bridge circuit. For small displacements the capacitance between a horizontal plate, above the surface, and the surface itself can be measured. Interferometric techniques can be used for very small displacements, such as occur in ripple tanks[47] (which are used as analogues for other forms of wave motion). Microwave interferometers have been used to track the movement of gun-tunnel pistons.

Automatic Data Recording and Digital Processing

Many of the measuring instruments described in the last three chapters have as output a voltage proportional to, or at least related to, the quantity being measured. Other instruments can be modified to give an electrical output: for instance, liquid-level manometers have been fitted with electrodes driven by a servo system so that they follow the surface level of the (electrically conducting) liquid. In the past, output voltages were recorded on chart recorders and then read off by eye for any further analysis to be done, but recording in digital form for subsequent entry to a computer is now much more common, even in small laboratories and in the field.

An "analogue" electrical signal V volts, say, can be converted into digital form by a digital voltmeter (DVM). One common type works by simultaneously switching on, at time $t = 0$, a pulse generator of frequency f and a "ramp" voltage αt. When $\alpha t = V$, the pulse generator is switched off, after emitting a total of Vf/α pulses, which are recorded by a pulse counter. The pulse counter reading is output (to "output", with the stress on the last syllable, is a verb frequently used in discussions of data processing) as a row of illuminated decimal digits and as a binary coded decimal (BCD) pattern of voltages on terminals at the back of the instrument.

Binary counting is almost universally used in computing because numbers can be represented by two digits, "on" and "off". It uses powers of two exactly as the decimal system uses powers of 10: for instance the decimal number 9 is written as 1001 ($1 \times 2^3 + 0 \times 2^2 + 0 \times 2^0 + 1 \times 2^0$): we see that any number from 0 to 9 can be represented by voltages on four terminals. To represent the decimal number 1234 in BCD as 0001/0010/0011/0100 requires a minimum of 16 terminals (more are usually employed to give a "parity" check, of which the simplest form is to add one more digit to make the number of binary "1" digits even, say). BCD is used for most computer input devices because it is necessary to pass from the pure decimal used by human beings to the pure binary used by computers: the pure binary for decimal 1234 is 10001101110, unhandy to write and remember. Octal (powers of 8) is more compact than binary and is used for some purposes in computers.

The BCD output voltages from the DVM or other instrument are fed to a punch drive unit which feeds pulses of current to a paper tape punch: the tape has at least five tracks (the above four binary digits plus parity) in each of which a hole can be punched to signify a binary "1", so that the decimal number 1234 would appear on the tape as:

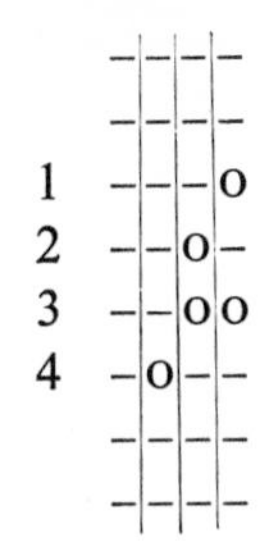

where "o" denotes a hole and "–" no hole. Whether one reads the tape with the least significant digit on the right, as in normal counting, or on the left, is a matter of preference. Eight-hole tape is now becoming standard, to permit a full set of decimal digits, letters and mathematical symbols to be written without the "case shift" symbols needed with smaller numbers of tracks. If the paper-tape punch is combined with an electric typewriter, information such as the date, the run number and so on can be put on the tape by hand, and a printed record of all the data on the tape is available for inspection. Similar facilities can be used with punched cards instead of paper tape, but whereas there are some advantages in using cards for computer programs that have to be altered from time to time, paper tape seems to be far more convenient for data handling. When data must be stored at a high rate or in large quantities, magnetic tape ("magtape") is used instead of paper tape, a binary "1" corresponding to a change of direction of magnetization of the tape: magtape is mainly intended for pure binary use and 7-track (6 binary digits plus parity) or 9-track (8 plus parity) tape is standard. Although the different codes and conventions used may seem complicated to the beginner, a lot of hard work has been needed to achieve something approaching standardization of computer input and output equipment.

Digital voltmeters, "analogue-to-digital converters" (which are fast, lower-accuracy DVMs converting directly into binary) and all the equipment for handling paper tape, cards or magtape, can be obtained commercially. The combination of equipment for fast data recording and a computer for fast data reduction permits a given experiment to be completed more quickly, but perhaps the real advance is that enormously complicated experiments can be done *and* the results presented in fully analysed form without all concerned becoming thoroughly bored with taking or analysing readings. It is possible to digitize a signal and record it on magnetic tape at 30,000 samples a second, about 10^5 times as fast as a human can do it, and then do arithmetic operations in a computer at a rate of 10^5 per second, 10^6 times faster than a human. It still needs a human to decide what data to record and how to interpret the results: automatic data recording and processing give him more time to do it.

Examples

1. Granted that strain gauges would have insufficient range for use on a permanent null-displacement balance and are therefore customarily used with sting balances, why are null-displacement balances so rarely used with sting-supported models in high-speed tunnels?

2. If neither strut supports nor a sting are acceptable for mounting a model (if, say, base pressures in transonic flow are of interest) what other methods could be used?

PLATE 1. Paraffin-mist filament-line picture of flow over a delta wing: angle of sweep 70 deg, angle of incidence 20 deg (see p. 145).

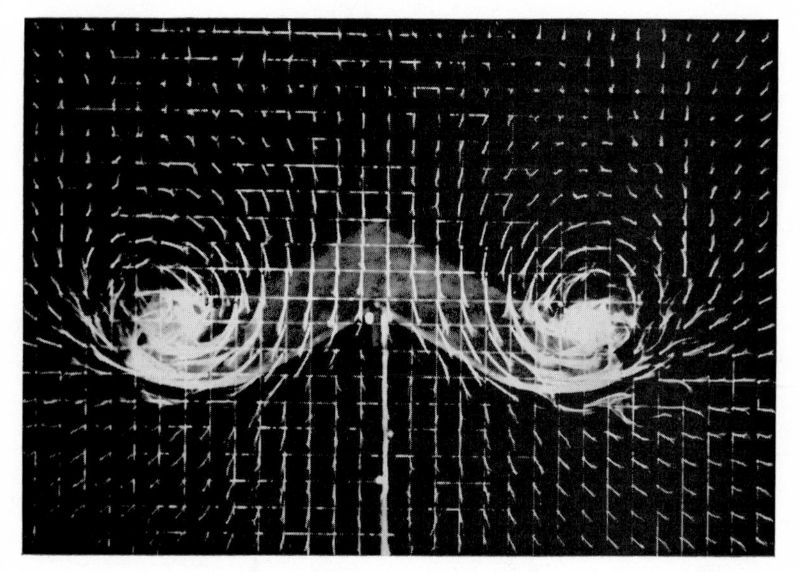

PLATE 2. Tuft-grid picture, looking upstream, of flow behind a swept wing: angle of sweep 60 deg, angle of incidence 30 deg (see p. 137).

PLATE 3. Surface rib-flow picture of flow round a "boundary layer fence" near the leading edge of a swept wing. Flow from top to bottom of picture. See Fig. 46 for a partial explanation.

PLATE 4. China-clay picture of longitudinal vortices in a wind-tunnel contraction boundary layer: three-quarter side view (see p. 149).

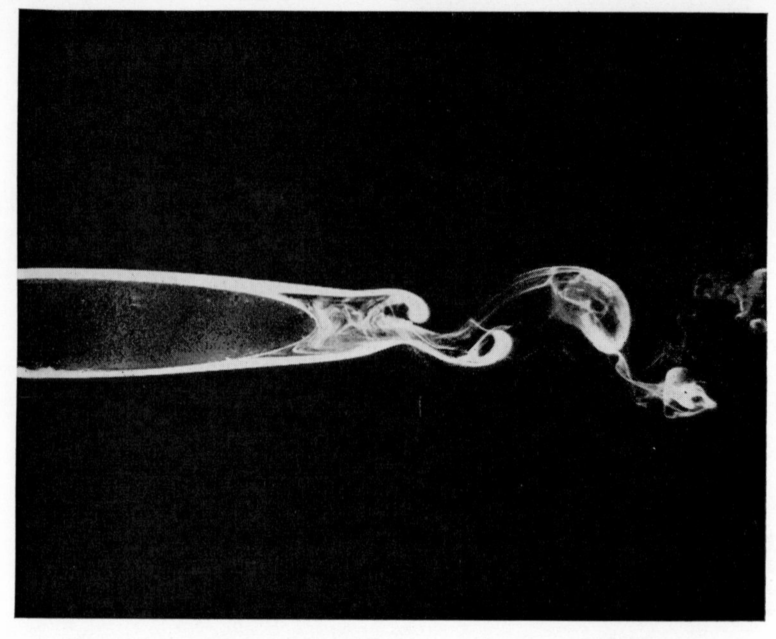

PLATE 5. Titanium tetrachloride vapour picture of separation, instability and transition of the wake of a two-dimensional elliptic cylinder (see p. 149).

PLATE 6. "Breaking" of internal waves in dye-filled thermocline shear layer; wave moving from right to left (see p. 183).

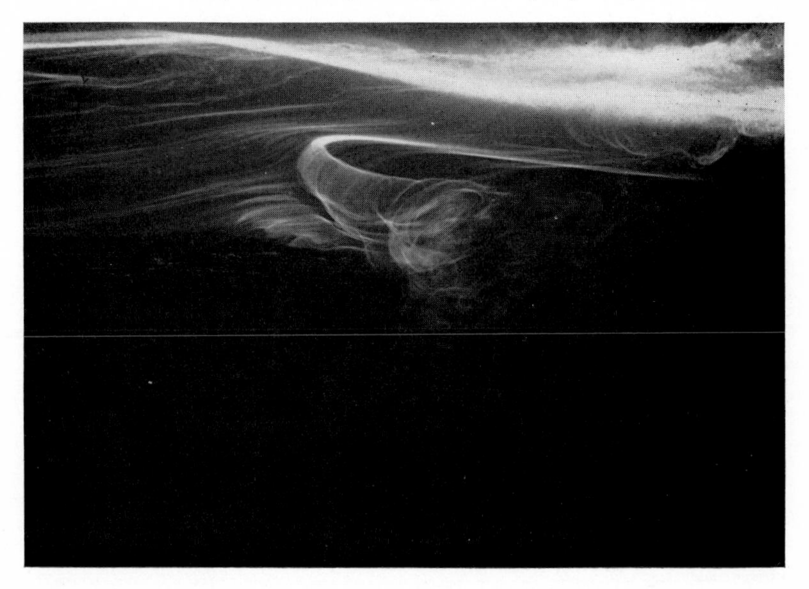

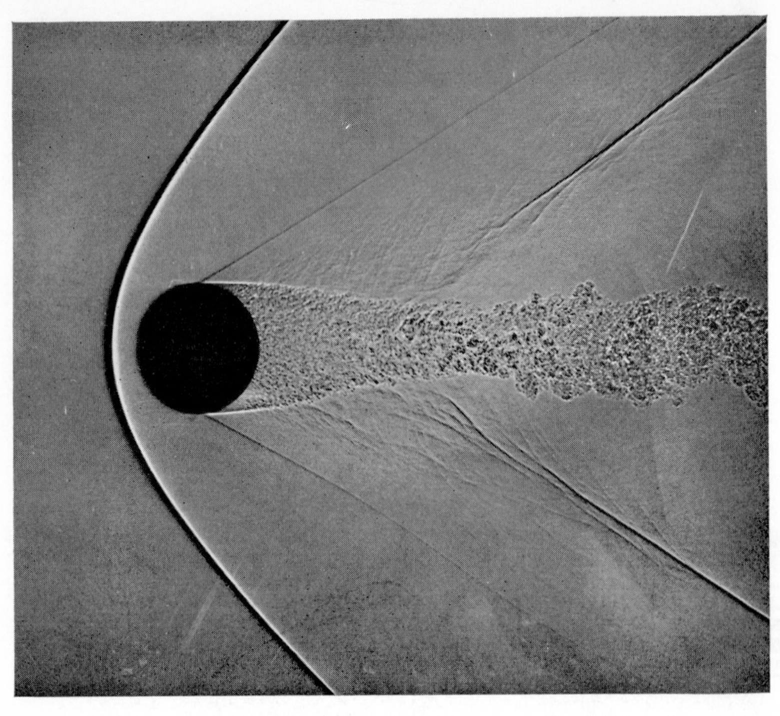

PLATE 7. Shadowgraph picture of shock waves and wake of a sphere in free flight. Mach number 1·6 (see p. 150).

PLATE 8. Hydrogen bubble picture of flow in a turbulent boundary layer. Wire spanwise $u_{\tau}y/\nu = 50$; flow from top to bottom of page.

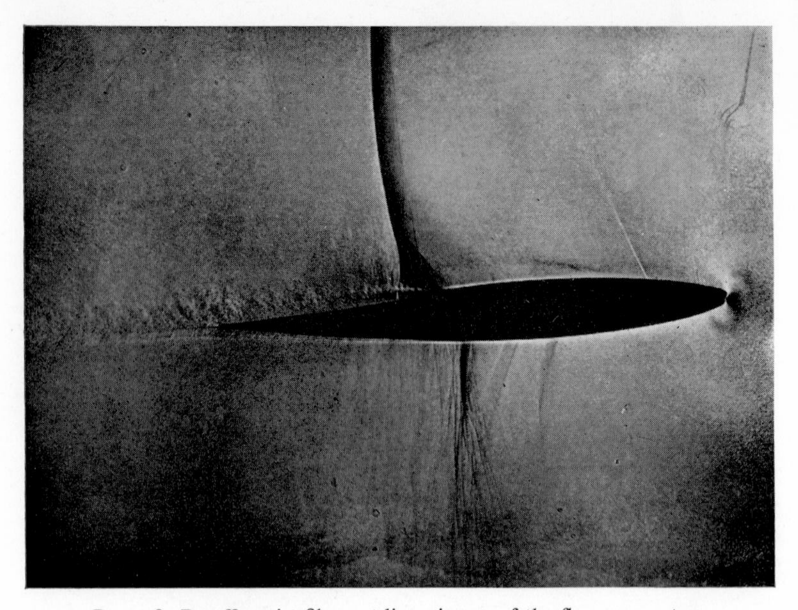

PLATE 9. Paraffin-mist filament-line picture of the flow over a two-dimensional jet-flap aerofoil (see p. 150).

CHAPTER 6

Flow Visualization

FLOW visualization is a very satisfying technique to use because the information it gives can often be assimilated at a glance without the need for tedious data reduction, and because it is possible to get an idea of the whole development of the flow, either in the boundary layer or in the external stream, from one observation. However, flow pictures may need considerable experience to interpret, and rarely give quantitative information of the sort normally required from experiments in fluid mechanics. Therefore, although elegant and useful experiments have been carried out by the use of flow visualization techniques alone, the methods to be described are most powerful when used together with techniques giving numerical results. The most frequent use of unsupported visual methods is in checks on the performance of the test rig. A review of visualization techniques used in liquids is given in ref. 48.

Flow visualization is capable of giving information about five main features of the flow, which we discuss before proceeding to describe the experimental techniques.

Streamlines of the External Stream
(Plates 1, 2 and 8)

This information is usually needed only when boundary-layer separation occurs, when jets are blown into the stream, or in highly three-dimensional flows. In other cases it is possible, with practice, to guess the position of flow streamlines to an accuracy nearly as good as that obtainable from tracer techniques: if really accurate information is needed (for instance to find the downwash induced at the tailplane position by the wing of an aircraft) then yawmeter observations must be made.

The two main methods of visualizing streamlines or flow directions are the particle or filament technique in which dye, smoke, foreign gas, heat, bubbles or solid particles are introduced into the stream at selected points, and the tuft or streamer technique. The

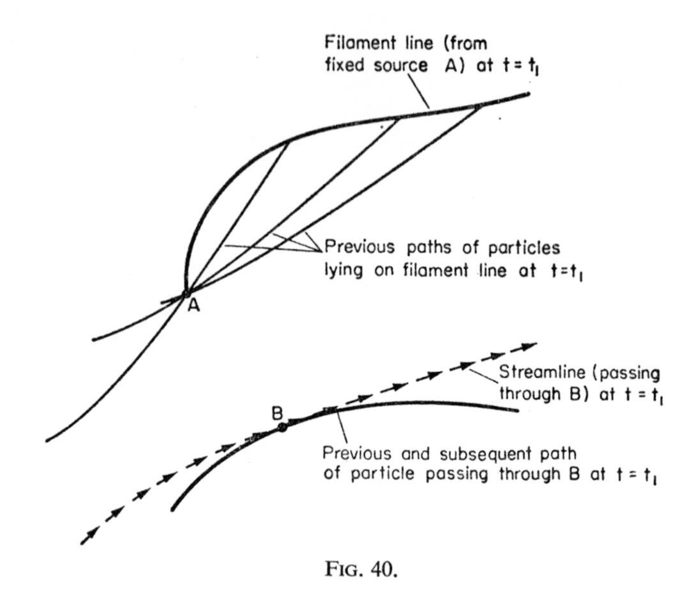

FIG. 40.

tuft method gives the flow direction at each tuft position but it is difficult to deduce complete streamline shapes: providing that the tufts follow the motion of an unsteady flow without appreciable lag they will still indicate instantaneous flow directions and a periodic flow can be observed with a stroboscope. However, the interpretation of tracer patterns in unsteady flow (and, in general, there is at most only one frame of reference in which a flow is steady) is more complicated, because one must distinguish between (Fig. 40):

streamlines, which are lines to which the *instantaneous* direction of motion at any point is tangential, so that fluid is not flowing across them at the instant considered;

filament lines, which are the *instantaneous* loci of all fluid particles which have passed through certain points, and can be made visible by smoke or dye injection at selected points;

and *particle paths*, which are the loci of certain fluid particles, and can be shown up by time-exposure photographs.

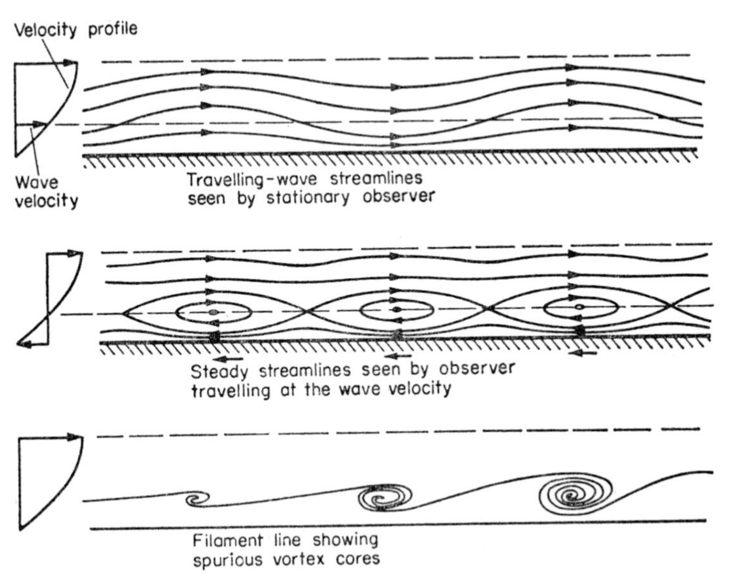

FIG. 41. Streamlines and filament lines in a shear layer with travelling instability waves.

As an example of the possible pitfalls, Hama[49] has pointed out that a filament line in a travelling wave flow, such as a shear layer with superimposed instability waves, will roll up into cores and give the impression that discrete vortices are being formed whereas in fact the streamlines are still sinusoidal. This may be seen by considering the streamlines as seen by an observer moving with the wave velocity c (see Fig. 41), the so-called "cat's eye" diagram.

The flow appears steady to this observer so that streamlines and particle paths coincide: fluid on the closed streamlines therefore moves in closed paths and filament lines will develop cores.

Particles can be used, like tufts, to show instantaneous flow directions by taking a short-exposure photograph, when the particle paths show up as short streaks, whose length is proportional to the local velocity.

Streamlines of the Surface Flow
(Plate 3)

In three-dimensional flow, streamlines in the slow-moving flow near the surface may be very different in direction from those in the external stream, since approximately the same pressure gradient acts on both and the slow-moving fluid has the smaller inertia. It can be seen from the equation of motion in the z direction (see p. 17) that if viscous forces are neglected typical radii of curvature will be proportional to the square of the local velocity, in order that the pressure gradient in the z direction shall be balanced by an apparent centrifugal force. An everyday example of this is the difference which may often be observed between the wind near ground level and that at cloud level, although the flow in this case is made more complicated by the apparent Coriolis force caused by the rotation of the Earth. Near positions of boundary-layer separation and obstacles on the surface and beneath vortices in the flow, the surface streamlines may become very complicated indeed. A particularly convoluted example is shown in Plate 3; the student should not be discouraged by failure to understand all the details of this photograph, and a key to the main features is given in Fig. 46. It is, of course, as a means of detecting the events leading up to separation that surface flow techniques are chiefly used. Smoke or tufts can be used near a surface in the same way as in the external stream, but a more convenient method is to smear oil, or an oil–dye mixture, on to the surface: in the absence of gravitational effects the oil will move slowly in the direction of the shear stress at the surface, that is, in the direction of the limiting streamlines at the surface, which show up as streaks in the film.

Surface Friction
(Plate 4)

The intensity of surface friction beneath a boundary layer increases more or less suddenly in the region of transition from laminar to turbulent flow, decreasing to zero and then becoming negative in regions of separation. These large changes in shear stress at the surface can be detected by the oil-film method, but a more positive indication can be obtained by making use of the close correspondence between surface shear stress and the rate of sublimation of a solid such as naphthalene or the rate of evaporation of a liquid film, as in the china-clay method. Changes in shear stress could also be detected by means of the reading of a Pitot tube placed at various positions on the surface: this would, however, be a tedious process, especially in three-dimensional flow, and the comparative simplicity of the evaporating-film methods gives a good example of the advantages of visual techniques when quantitative measurements of pressure or velocity are not required. Except in experiments on the artificial maintenance of laminar flow, the chief reason for interest in the position of transition is as a check on the performance of trip wires or distributed roughness used to precipitate transition: this in its turn is a good example of the use of visual techniques to make checks on the performance of the test rig.

Many methods can be used for separation indication, relying on the departure of the external flow streamlines from the surface (Plate 5), or the onset of backflow, and the adaptation of the methods described below will generally be obvious. A unique method of obtaining a permanent record of separation position is to smear the surface with oil or grease and to blow fine powder through a tube into the separated flow region: powder will eventually adhere to the surface everywhere behind the separation line at which reversed flow begins. A recent development[50] is the use of a "smoke" composed of paraffin wax particles condensed from vapour; these particles adhere naturally to an untreated surface. These bulk-injection methods are particularly useful in three-dimensional flow (though separation is nearly always three-dimensional to some extent even in a nominally two-dimensional flow).

Extent of Turbulent Mixing Regions
(Plates 5, 6)

If smoke or other contaminant is introduced in sufficient quantity into a turbulent region it will be uniformly diffused over the whole thickness of the mixing region further downstream and will thus show up the boundary of the turbulent fluid.

Shock Wave or Expansion Fan Position
(Plates 7, 9)

The intersections of shock waves or expansion fans with the surface of a body can be deduced from pressure-plotting measurements, with the reservation that the exact position of a sudden pressure change cannot be found from measurements at fixed static pressure tappings, but it is much easier to use visual techniques for showing up the corresponding density changes, and in addition the wave pattern in the external stream can be observed. The methods to be described here, the shadowgraph, schlieren and interferometer methods, all rely on the change in refractive index or light propagation speed which results from a change in gas density, but they differ in their response to density changes. The interferometer is best suited to quantitative measurement of density and is less often used purely for visualization purposes than the other methods. The greatest drawback to these methods is the difficulty of applying them to three-dimensional flow, as will appear from the discussion below: none of the adaptations so far made for examining density changes at one point only along the light path has proved really satisfactory.

Virtually all high-speed tunnels have permanent installations of one or other of these optical methods which are used as a matter of course for keeping a check on the flow behaviour. The state of the boundary layers on the model, as well as the position of shock waves, can be observed if the flow speed and density are high enough to show up the smaller density changes that occur in laminar and turbulent shear layers. Unsteady flows can usually be observed by spark photography. The greatest difficulty in interpreting the pictures is in deciding whereabouts on the span of the model the observed effects are occurring: even in nominally two-dimensional flow, the influence of the tunnel wall boundary layer on the flow near the ends of the model may be important.

Flow Visualization Techniques and Interpretation[51]

Tufts

The usual material for tufts for use in air is tufted nylon yarn which is spun so that when unstretched it contracts to a fraction of its stretched length because of the crinkling of individual fibres, and then looks like very soft knitting wool. It can be used down to speeds of one or two m s^{-1} in air before stiffness and gravitation affect the results: at speeds of more than about 30 m s^{-1} in air it starts to stretch appreciably under the action of air loads until at the highest speeds reached in "low-speed" tunnels it resembles ordinary thread, which is an alternative material for work at such speeds. If tufts are to be distributed all over a model they can be glued to the surface, but as one frequently wishes to change the position of odd tufts applied *ad hoc*, cellulose tape can be used instead. The tuft length should clearly be a small fraction of the expected radius of curvature of the streamlines: the stretching properties of tufted nylon are sometimes a disadvantage in this respect. Even the smallest and shortest tufts may produce transition of a laminar boundary layer and the application of too large a number may so thicken the boundary layer as to cause premature separation: if there are doubts about the effect of tufts on the boundary layer they should be applied progressively starting at the trailing edge, to see if the flow near the trailing edge is affected by the addition of tufts further upstream.

If tufts are to be used to examine the external flow they may be supported on wires: complete grids of wires normal to the flow have been used to show up trailing vortices (Plate 2).

If tufts are short and do not interfere with the flow the interpretation of their attitude is straightforward. *Motion* of surface tufts in a nominally steady flow may be taken to mean that the boundary layer has become turbulent, though large-scale unsteadiness of the tunnel flow may cause similar effects. Violent motion of the tufts, or a tendency to lift from the surface and point upstream, indicate separation.

Smoke or Dye

Much of the ingenuity required to use smoke or (more usually) condensed vapour in gases, or dye in liquids, is directed towards introducing it into the stream without disturbing either the flow or the operators. When only small quantities of "smoke" are required the Preston–Sweeting paraffin mist generator (Fig. 42) or its variants are adequate, easy to start after a little practice, and capable of running indefinitely without attention except for draining the condensed paraffin from the pipes leading to the model or the smoke injector. When large quantities are needed, genuine smoke, obtained as the product of combustion, may be used: a variety of suitable pyrotechnic devices is commercially available, and many home-made smoke generators have been described.

The effluent from these devices may contain carbon monoxide or other toxic compounds, and, in addition, products of combustion are likely to clog pipes and corrode metal. (Many of the compounds suggested for flow visualization in general are toxic and therefore unsuitable, though on several occasions this has only been realized indirectly after prolonged use without ill effects.) A particularly brilliant white smoke is obtained from the reaction of titanium tetrachloride with atmospheric water vapour, producing a mist of titanium oxide particles and hydrochloric acid:

this compound is useful for dabbing on to a surface (Plate 5) but, like particle smokes in general, is not very suitable for introducing filaments into the external flow, and its other disadvantages will be obvious. In water, potassium permanganate solution may be used as a dye, though it again is rather corrosive: fluorescein or eosin are also suitable.

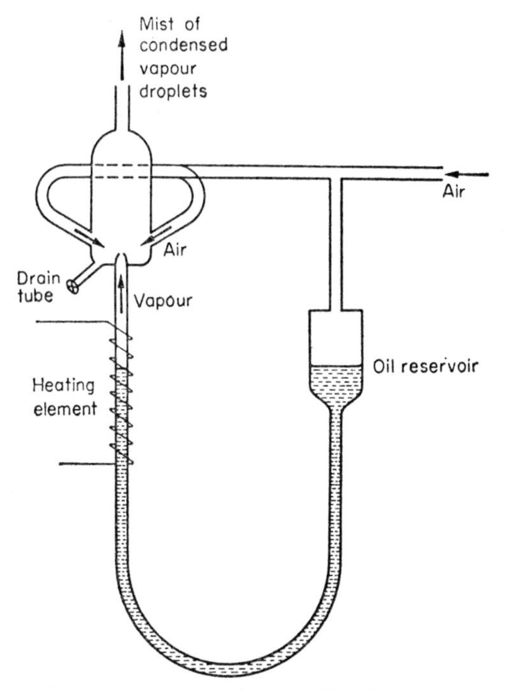

FIG. 42. Preston–Sweeting paraffin mist generator.

The production of filament lines in the external flow is very difficult, as it is necessary to prevent the wake of the smoke or dye injector from becoming turbulent. This technique is virtually confined to low-speed liquid flows and special smoke tunnels which run so slowly that the Reynolds number of the injector wake is below the critical: this implies that the Reynolds number of the model is also low. The smoke or dye is emitted from short pipes projecting from the trailing edge of a streamlined aerofoil. The aerofoil may have boundary-layer suction applied to its surface to suppress the wake, or be placed in a sudden contraction so that the velocity gradient over its surface is always positive (again tending to thin the boundary layer and wake), or be placed immediately upstream of a screen with the tubes projecting through the screen, so that the velocity defect in the wake is reduced.

Smoke tunnels are usually of open-circuit design to prevent the accumulation of smoke in the airstream: the notable disadvantage of this arrangement is that the smoke accumulates in the tunnel room instead, because the sensitivity to draughts of these very low-speed tunnels generally precludes direct discharge to the open air. It would seem possible to use a closed-circuit tunnel for smoke tests if the air in the tunnel could be blown out into the open air at intervals: no smoke need enter the tunnel room and the tunnel would be insensitive to draughts.

Photography of smoke or dye flows needs care with lighting: the best arrangement seems to be to light the flow field from above and below and to photograph from the side, taking care that the windows are not illuminated and that no deep shadows occur. Paraffin mist usually appears slightly blue in colour, and an improvement in contrast can be obtained by using blue-sensitive plates or film. The arrangements used in the N.P.L. smoke tunnel, in which Plates 1 and 5 were taken, are shown in Fig. 43.

In the smoke-screen technique, large quantities of smoke are released into the tunnel, usually by means of smoke bombs in the settling chamber, and illuminated in a single plane normal to the flow by shining a beam of light through a slit into the tunnel. Cross-sections of vortices and separated-flow regions can be seen, and the method is generally more convenient to use than the tuft-grid method which gives very similar information. A related "vapour screen" technique is used in high-speed tunnels: humid air is deliberately used so that condensation occurs in the working section: the vapour is then illuminated as above.

Solid-particle tracer techniques are not easy to apply in air although aluminium flakes, soap bubbles, and the white flakes produced by the heating of metaldehyde ("meta fuel") have been used. The measurement of wind velocity by means of free-flying balloons strictly falls into this category, and is probably used more frequently than all other flow visualization techniques put together. In water and other liquids a wide variety of particles has been used. Ideally the density of the particles should be the same as that of the liquid: polystyrene has a density about 1·03 times that of water

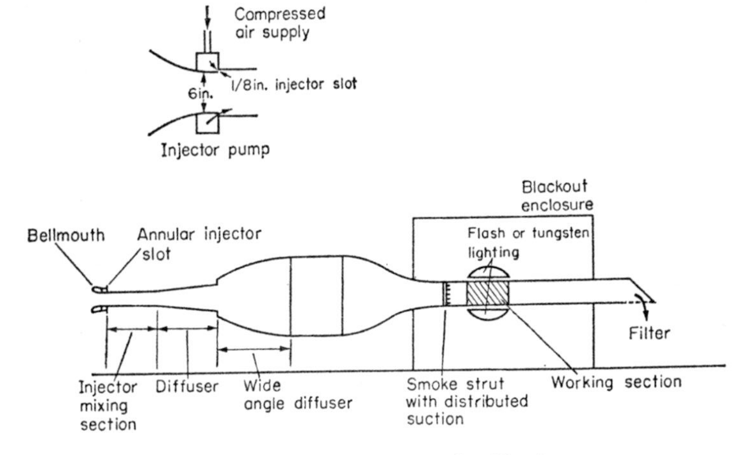

FIG. 43. N.P.L. Smoke tunnel—side view.

and is available in small beads which are very suitable for water tunnel use. Particles of *dye* are sometimes used: ref. 52 describes how falling dye particles were used to visualize currents in the surface layer of the ocean.

The optical techniques used for visualizing shock waves can be used to observe the path of foreign or heated fluid as they will show up changes in refractive index, however caused, either in liquids[53] or in gases. The shadowgraph was used by Townend in his heated-wire method[8] (analogous to the hydrogen bubble wire) as long ago as 1930, and a more recent application, using volatile liquids in conjunction with a schlieren system is described by Pierce.[54] Freon-12, the refrigerant, has one of the highest refractive indices of any gas, about 1·0010, and can be introduced in exactly the same way as smoke. Vertical displacements of a density stratified fluid also produce density changes at a given height: internal wave patterns can be seen on a schlieren system[55].

Line Tracers

The hydrogen bubble technique[56] is now in common use in water, especially for turbulence studies. A potential of a few tens of volts is applied to a platinum wire, typically about 0·003 cm diameter, and hydrogen bubbles, whose diameter is of the same order as that of the wire, are generated by electrolysis. It is sometimes necessary to add a little electrolyte to the water, but the author has successfully used the rather "hard" tap water of the London area. The second electrode should also be made of platinum to stop the unwanted transfer of metal ions between the electrodes. The polarity of the voltage should be reversed occasionally to clean the wire (the oxygen bubbles then produced are larger, fewer in number and therefore less suitable for flow visualization). Plate 8 shows an arrangement for quantitative or semi-quantitative measurements, in which the applied voltage is pulsed, and portions of the wire covered with insulating paint, so that bubbles are produced in rectangular patches whose size, shape and separation can be measured to deduce velocities and velocity gradients. Individual spanwise lines of bubbles can be used to show velocity profiles directly.

Another method of producing a line of tracer normal to the flow direction is the flash photolysis technique, in which a narrow beam of light is used to catalyse a chemical colour change in the liquid. The reaction between ferrous and ferricyanide ions used in the "blueprint" process has been used but is rather slow, and Popovich and Hummel[57] have used a solution in ethyl alcohol of a dinitro-benzyl-pyridine, which changes very rapidly from colourless to blue on irradiation because of a change in molecular structure (reversion to the original structure occurs within a few milliseconds, which is, on balance, a disadvantage).

Oil Films

Unpigmented oil can be used to obtain surface flow pictures: heavy oils suitable for high-speed wind tunnel use are naturally fluorescent and good contrast is obtainable by viewing the film in ultraviolet light, while at lower speeds light oils can be used with a fluorescent additive. An alternative is to use a mixture of oil and titanium oxide or other pigment. The viscosity of the oil or the proportions of the mixtures must be varied to suit the tunnel speed, but at best very clear patterns of streaks (Plate 3), following the surface streamlines, can be obtained. The reason for the appearance of streaks seems to be that any small hump that does appear will tend to protect the oil in its lee from being blown downstream, so that the hump will collect more oil by a snowball effect as it moves downstream. Low-viscosity oil tends to form U-shaped waves which deposit streaks at their streamwise edges. The disadvantage of applying oil mixtures to the surface of a model is that pressure-plotting holes may be clogged: usually the holes are deliberately blocked with some easily soluble compound before flow visualization studies are made.

If the flow patterns are complicated and include regions of reversed flow some care may be needed in interpreting the patterns of streaks as of course the *sign* of the wall shear stress is not indicated. It is advisable to apply the oil in as distinctive a pattern as possible so that swabbing marks in stagnant regions are not taken for evidence of strange flow phenomena: a fairly even distribution can be obtained by using a paint roller. Care must also be taken that the oil does not follow machining marks on the surface: the writer recalls being greatly mystified by evidence of pronounced secondary flow on part of the floor of a rectangular duct, which proved to be due to machining marks although the surface was quite adequately smooth for other purposes.

China Clay

The best of the evaporation techniques[8] for indication of transition is the china-clay technique. The surface of the model is sprayed with a thin coat of a suspension of china clay in clear cellulose lacquer and smoothed down carefully when dry. It is then sprayed with a liquid of nearly the same refractive index as the china clay itself (1·56) so that internal reflections in the crystals are prevented and the layer becomes transparent. The china clay is *not* dissolved or suspended in the liquid: the general effect is the same as if it were, but the solution would then tend to be blown downstream like an oil film and the results might be confusing. When the tunnel is run the liquid evaporates more quickly from the surface where the flow is turbulent and the film again appears white in these regions. The liquid can be chosen to give a reasonable evaporating time: methyl salicylate (oil of wintergreen) has a refractive index of 1·54 and is convenient for most wind tunnel work at low speeds, but many other liquids have been used for different speed ranges. Plate 4 was taken using acetone (which probably dissolved some of the china-clay lacquer): the speed in the wide part of the contraction was only about 3 m s^{-1}. The wedges of apparent turbulence in this picture are caused not by isolated roughness elements but by Taylor–Görtler centrifugal instability[3] which results in the formation of longitudinal vortices in the boundary layer on concave surfaces.

Shadowgraph or Direct-shadow Method

If a parallel beam of light travelling in the z-direction passes through a region having a gradient of refractive index in the y-direction each ray will be deflected through an angle

$$\theta = \int \frac{1}{\mu} \frac{\partial \mu}{\partial y} \, \mathrm{d}z \tag{29}$$

where μ is the refractive index: $(\mu - 1)$ is proportional to ρ for a given fluid and equal to 0·00029 for air at standard density. If the beam now falls on a screen the illumination will be increased where the rays are converged, that is where $\partial\theta/\partial y$ is negative or $\partial^2\mu/\partial y^2$ is negative, and decreased where $\partial^2\mu/\partial y^2$ is positive. This effect is responsible for the shadows cast by hot air rising from fires or radiators. For wind tunnel work the parallel beam is produced by a point source and a converging lens or mirror, and passed through the working section parallel to the span of the model under observation. Shock waves then appear on the screen as two adjacent bands, one dark and one light (Plates 7, 9) corresponding to the sudden increase in density gradient at the front of the shock and the sudden decrease in gradient at the rear (see example 4). Clearly, expansion fans which, although they may have total density changes as large as those occurring in shock waves, have much smaller density gradients and very much smaller rates of change of density gradient, will not show up well in shadowgraph pictures. An interesting advantage of the shadowgraph is that transition can be detected easily, because the position of maximum rate of change of density gradient is further from the surface in a laminar boundary layer than in a turbulent boundary layer of similar momentum thickness, so that the bright line marking this position in the flow dips toward the surface as transition occurs.

Schlieren Method

It should be noted that schlieren is not a man; only a German plural noun meaning "streaks", especially the streaks which appear on looking through imperfect glass, and the name is more properly used to describe several methods including the shadowgraph. The method normally used in aerodynamics is the Töpler schlieren method (A. Töpler, fl. 1870). Referring to the expression for angular deflection, eqn. (29), we see that if the parallel beam is brought to a focus after passing through a region of varying refractive index (Fig. 44), the linear deflection of a particular ray at the focus will be proportional to the density gradient. If a knife edge or a graded filter is now inserted at the focus so as to cut off a fraction of the image of the light source, the intensity of illumination of a screen placed beyond the focus will be uniformly reduced by the same fraction, except that regions illuminated by the deflected rays will be lighter or darker accordingly as the ray is deflected away from, or on to, the knife edge or the more opaque side of the filter. The change of illumination is therefore a function of the density gradient, and not of the rate of change of density gradient as in the shadowgraph method. The light source is usually a slit parallel to the knife edge, so that the image at the

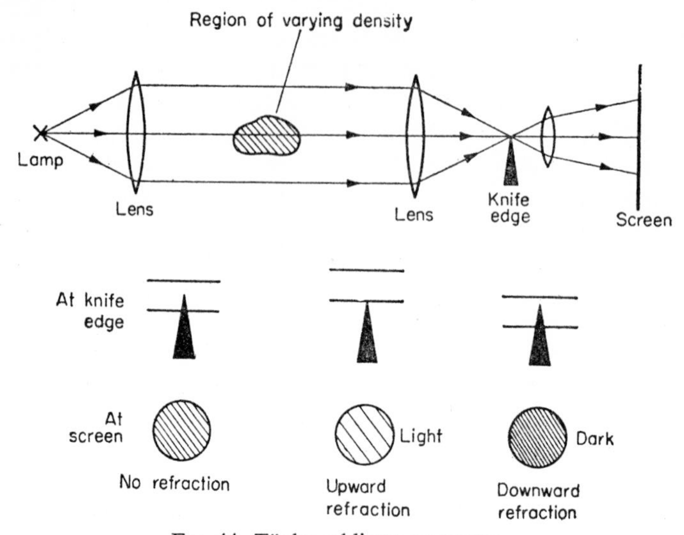

FIG. 44. Töpler schlieren apparatus.

focus is also a slit of finite height; each ray (unless deflected) occupies the whole of the image, so that a particular part of the image does *not* correspond to a particular part of the field. The sensitivity of the system (percentage change in screen illumination for a given change in density gradient) is inversely proportional to the height of the unobstructed part of the image at the knife edge. The maximum density gradient that can be represented by a directly proportional change in illumination is clearly that which deflects the image of a light ray so that it is entirely unobstructed or entirely cut off.

Because of its ability to show up more gradual changes of density than are visible by the shadowgraph method, the schlieren system is the more commonly used for flow visualization in wind tunnels. In practice mirrors are used instead of the large lenses shown in Fig. 44: as in telescope manufacture, mirrors are cheaper and their optical quality better. The schlieren system is only sensitive to density gradients in a direction perpendicular to the knife edge, and it is usual to mark upon schlieren photographs the orientation of the knife edge to help in interpreting the picture. Shock waves appear as single lines, dark or light according to the orientation of the knife edge, and expansions appear as extended dark or light areas. In Plate 9 an expansion occurs round the leading edge of the aerofoil, the deflection of the light rays causing an apparent distortion of the profile, and the flow over the front half of the aerofoil is supersonic, as can be seen from the weak inclined shock waves. The boundary layers over the front half are thin and not very clearly seen. The shock waves which close the supersonic regions cause a considerable thickening of the boundary layer on the lower surface and a separation from the upper surface, which since the aerofoil is lifting, has a lower static pressure, higher local Mach number and therefore a stronger shock wave. The λ-shaped shock on the top surface is typical of the pattern produced by a shock-wave boundary-layer interaction.

Interferometer[58]

A direct response to density changes is given by the interferometer, Fig. 45, which depends on the interference fringes formed on the recombination of two light rays from the same monochromatic source which have taken different times to make the journey. If the two path lengths are the same, interference fringes may be produced by a region of different light propagation speed in one of the paths. The Mach–Zehnder interferometer uses a half-silvered plate to split the single beam from the light source into two beams, one of which by-passes the working section, and which are recombined at a second half-silvered plate before falling on a screen. The light paths are adjusted with no airflow disturbance to produce a uniform and parallel set of interference fringes on the screen, or possibly so that both beams are exactly in phase at the screen, giving uniform illumination. In the first case the fringe spacing will change, when the tunnel is run with the model installed, by an amount proportional to the phase change introduced by the disturbance at any point, which is in turn proportional to the change of fluid density integrated along the light

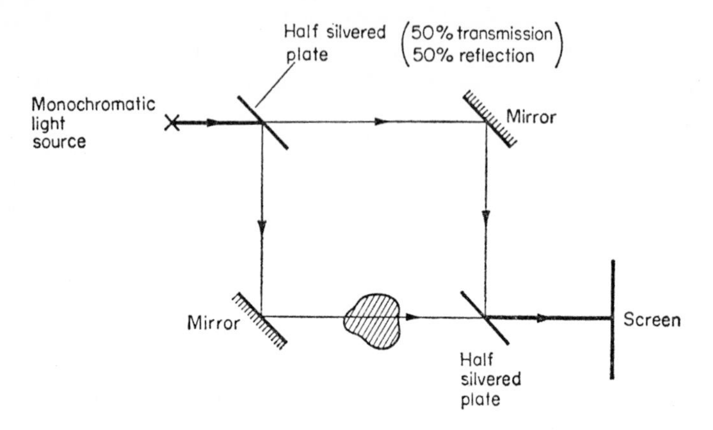

FIG. 45. Mach–Zehnder interferometer.

path. In the second case, which is of course more difficult to set up, fringes will appear only in the regions where the density is altered by the disturbance, and will indicate contours of constant density. In either case, the interferometer must be calibrated, either from theory or by comparison with pressure measurements, and will then give absolute values of density.

The use of a laser to provide a light source that is exactly monochromatic and coherent (uniform phase) with high intensity, a parallel beam and a small aperture has made it much easier to set up an interferometer. The greatest benefit is that if the source is exactly monochromatic and coherent the light paths need not be the same length. The first beam splitter shown in Fig. 45 is required merely to by-pass part of the beam round the test rig, and the design of the second is greatly simplified (for instance a simple mirror can be used: the beams are arranged to intersect close to the surface of the mirror and interference occurs between the intersection and its image). A discussion of the adaptation of existing types of interferometer for use with laser light sources is given by Tanner[59].

Before the advent of lasers interferometers were used in preference to the simpler shadowgraph or schlieren system only when quantitative density measurements were needed and could not be obtained by other means, as in the study of molecular relaxation (see p. 32) behind moving shock waves, but they may become more popular in the future for flow visualization with the added bonus of quantitative density measurements if required.

Examples

1. Streamlines and filament lines coincide in steady flow past a stationary body. Do they coincide if the body is moving through stationary air and seen by a stationary observer?
2. What are the objections to the use of solid particles in air?
3. Does the white area near the narrow end of the contraction shown in Plate 4 from which the china-clay "solvent" has evaporated necessarily imply that the flow in this part of the contraction is turbulent?
4. Why do shock waves of almost infinitesimal thickness show up in shadowgraph pictures as two distinguishable bands?

CHAPTER 7

The Planning and Reporting of Experiments

THE experimental techniques used in the three divisions of fluid mechanics mentioned in Chapter 1, research, project testing and prototype testing, are similar and the planning and execution of the experiments may be much the same. Usually, the results of prototype tests do not require explanation by the experimenter but are submitted to the design staff in the form of undigested tables and graphs. The project test worker will probably have to modify the test model until the desired results are obtained, while the research worker must try to explain his results in terms of the phenomena occurring in the flow, or to compare them with theory. The physical insight necessary to explain results or improve performance cannot be taught: furthermore it is difficult to acquire it by reading other people's reports, since reports usually give an account of what was done without explaining the writer's mental processes or the route by which success was attained. Most experienced workers would claim that experimental fluid mechanics is merely a matter of the application of common sense, but then common sense is the fruit of knowledge and experience. All we can do here is to discuss in general terms the way in which the experimental techniques described in the other chapters are used, and to illustrate some precepts on the planning and execution of experiments by reference to a few published reports of typical investigations.

Planning of Experiments

Most leaders of experimental groups will have to develop or arrange a great deal of their apparatus themselves, though comparatively few will be directly associated with the design of large tunnels and other major facilities. In many cases the basic equipment like force balances, probes, traverse gear and manometers will be the same in each experiment, but there is inevitably some change to be made in the apparatus and its layout for each experiment, and, periodically, advances in experimental techniques or the ravages of time will make it necessary to replace some of the basic equipment. The experimenter is, or should be, partly responsible for the design of test models. It is almost invariably more economical in time and money to buy apparatus, if it is available commercially, rather than to make it, and unless the experimental group has a particularly comprehensive design office and workshop service it may be better to place orders with commercial firms for the development of the more complicated items of specialized apparatus. In all cases, however, the experimenter must learn enough about his apparatus to choose the right item from a commercial range or to give a coherent account of his wants to a drawing office designer or workshop mechanic: experimental fluid mechanics therefore embraces the rudiments of most of the branches of engineering, even including civil engineering where the construction of large test rigs is concerned.

It is, of course, possible to do good work with poor apparatus, but the work can be done better or more quickly if the experimenter does not have to spend time in nursing inadequate or temperamental equipment: one must try to strike a balance between the expenditure of time and the expenditure of money. Occasional periods of developing apparatus may make a welcome change from the collection or interpretation of data, but the experimenter should beware of going to the opposite extreme and spending too much time in the development of facilities and not enough on experimental work, much as development work may appeal to the latent engineer in every scientist. Clearly, the amount of time which must be spent in the development of apparatus will differ greatly between different types of experimental work, and will be greater for work of a pioneering nature than for investigations in a well-established field. It would save a great deal of duplication of effort if experimenters would publish details of the non-standard apparatus and techniques used by them.

Once the general purpose of the experiment has been decided the next stage in planning is to design the model or apparatus to be tested, paying due attention to the likely test programme. Let us suppose for example that a complete model of a projected or prototype aircraft is to be tested in a wind tunnel to find the effect of incidence, flap angle and so forth on the lift, drag and pitching moment. The model size must be chosen to suit the tunnel available (Chapter 2) so that the interference corrections (Chapter 5) shall not be excessive, and care must be taken to ensure that the test Reynolds number shall be large enough for the results to be applicable, with or without correction, to full scale: the model must be large enough for trip wires to be used to precipitate transition in the boundary layers, if necessary, without excessively disturbing the further development of the layers (Chapter 1, p. 28). Tests involving control surfaces are particularly susceptible to scale effect, because the ratio of the control surface chord to the boundary layer thickness must be large if the effectiveness of the control surface is not to be reduced by its immersion in a thick layer of slow-moving air. It may be possible to double the Reynolds number by using a half model, with one wing and one side of the fuselage mounted perpendicular to the tunnel floor or wall, on the argument that any stream surface, in particular a plane of symmetry, can be replaced by a solid surface if the flow is essentially inviscid: heed must be taken of this last proviso.

The range of movement of the control surfaces and any other movable parts of the model must be decided, and arrangements made for measuring their deflection and, if necessary, for adjusting their position without stopping the tunnel. The location of static pressure tapping holes must be decided with an eye to constructional difficulties: for instance tappings near a sharp edge, however desirable aerodynamically, are difficult to install.

Workshop staff can make or mar an experiment just as surely as the experimenter: it is important that they should have a good idea of what the apparatus is to be used for, both in order to maintain their interest and so that they can use their experience and discretion in deciding manufacturing tolerances without the expense and delay of producing detailed working drawings for "one-off" jobs. Time can be saved if the experimenter realizes, or if the workshop staff tell him, that a desirable feature could be incorporated in a model only at the cost of a disproportionate amount of extra work. Again, many pieces of apparatus can be made to very wide tolerances with the exception of one or two dimensions for which great precision is required, and if this is understood by the workshop staff they can save machining time.

The Experimental Programme

The number and range of variables to be tested must next be decided with reference to the capacity of the available measuring systems: the capacity of the force balance may restrict, say, the force that can be measured, and so limit the size of the model. This implies that rough estimates should be made of the likely ranges of dependent variables, whether or not such estimates are required for stressing the model. These estimates must also be made so that test values of the independent variables, such as incidence, can be spaced at suitable intervals: for instance, readings should be taken at small intervals of incidence in the neighbourhood of the stall. The spacing of readings can usually be decided finally while the test is actually in progress, providing that the significance of the results is apparent to the people who are taking the readings: the chief experimenter must decide whether his time is better spent in carefully planning the experiment or in constantly supervising the actual testing. Some large research establishments have arrangements for on-line computing so that the processed results can be inspected while the test is in progress, but this convenience is not available to many.

It will usually be necessary to estimate the likely duration of the tests, multiply the estimate by a fairly large factor of uncertainty, and arrange for the test rig to be available for this period, commencing some time after the delivery of the model so as to allow time for mechanical tests. Havoc will result if the apparatus arrives late, and requires modification, or if insufficient time is allowed for the test programme. Experimental science being what it is, it is as well to arrange the tests in the order that will suffer least from truncation due to delays, while still permitting the early results to be used as a guide in planning the rest of the tests.

When the apparatus is installed checks should be made to see that it is performing properly. Leak testing of the manometer connections and measurements of the balance zero readings should be carried out at this time. The first runs should be devoted to making sure that the flow does not differ wildly from that expected, that the required range of variables can be covered, and that the trip wires or roughness used to precipitate transition to turbulence in the boundary layers are in fact causing transition. This is where flow visualization methods can be used to advantage.

Methods of operating wind tunnels and test rigs are so varied that any general comments would be rather trivial. Strictly it is necessary to operate the test rig at constant Reynolds number, Mach number, etc., by making allowances for changes in fluid properties from day to day or even during a run, but these precautions are often relaxed and the usual procedure is to make complete runs at constant dynamic pressure (or total pressure and Mach number), so that manometer and balance readings remain constant, but to vary the dynamic pressure from day to day to keep the Reynolds number or Mach number constant if rapid changes of the non-dimensional coefficients with either of these parameters are expected.

Once the experiment has started, the results should be worked out into non-dimensional form as soon as possible: alternatively rough graphs of the raw data ("raw" in this sense being the opposite of "processed") should be plotted as the tests proceed. No general advice can be given about what to do if the readings seem to be nonsensical, but it is as well to start by checking the apparatus before seeking an aerodynamical explanation.

Experimental Errors

These fall into two classes.

Random errors are the discrepancies which occur between repeats of the same readings with the same apparatus, and can be allowed for by repeating the readings many times and making a statistical analysis of the results to establish the most probable value and standard deviation of the reading. If, as is usual, a test consists of a number of readings of the same dependent variable at closely spaced values of the independent variable some idea will at once be gained of the short-term repeatability of the results, and unless there is some reason for suspecting long-term repeatability it will be sufficient to repeat a few key runs once or twice to see if the discrepancies lie within the hoped-for accuracy of the experiment: these repeat runs should be made after as long an interval as possible.

Consistent or systematic errors are caused by the failure of the measuring instruments to respond exactly to the quantities they are supposed to measure. Failure to reproduce the desired flow exactly, because of scale effect or the effect of the boundaries of the test rig on the flow field, is probably best thought of as a limitation rather than an error, leaving the term "error" to describe a discrepancy between the actual result and the true value of what one sets out to measure in the laboratory. If the experiment is done only once, there is no way of knowing whether errors are consistent or random or indeed whether any errors have arisen at all: for instance, a mis-calibration of the balance used for force measurements could result in apparently consistent errors in all force readings until the calibration was checked. It is very difficult to estimate or allow for consistent errors in the measuring instruments: errors such as those caused by the inaccuracy of Pitot or static tubes or hot wires in a highly turbulent flow (Chapter 3) cannot be assessed without a great deal of extra work which may not be justified by the accuracy required.

Overall Accuracy

The figure generally quoted for the best accuracy required of experiments in fluid mechanics is 1 per cent, but this is to be regarded as a target rather than a necessity. Overall force measurements, pressure coefficients and even boundary-layer thicknesses can certainly be repeated to this order of accuracy in conventional test rigs but few people would claim that full-scale conditions could be simulated by models to such a high accuracy. At the other extreme the scatter of the earlier heat transfer measurements in shock tunnels was about 20 per cent, and few tunnel results in the transonic range can be relied on to represent the flow of an infinite stream past the model to much better than 5 per cent.

If one is attempting to measure the force coefficient of a body at a given incidence to an accuracy of 1 per cent, then, even assuming that the model is accurately made, the errors caused by the balance, the incidence inclinometer, the non-uniformity of the stream, and the manometer or other gauge by which the reference dynamic pressure is measured, must combine to be less than 1 per cent. If there is no correlation between the different sources of error, the final probable (root mean-square) error will be the square root of the sum of the squares of the individual errors, so that each of the four sources of error mentioned above could contribute $\frac{1}{2}$ per cent before the total probable error became 1 per cent.

Sometimes the correlation between different sources of error can be used to make the errors cancel out: if, for instance, the balance is of the strain-gauge type and the readings are being recorded by a self-balancing potentiometer chart recorder, the same source of excitation voltage should be used for the strain gauges and the potentiometer slide wire. Again, in experiments where a large number of pressure measurements are being made at different positions in the flow it is usual to apply the reference pressure difference to the same manometer: pressure coefficients are then obtained as the ratio of two liquid heads without worrying about the inclination of the manometer tubes or the density of the gauging liquid. These will be recognized as applications of the principle of non-dimensional presentation of results to facilitate the comparison of tests made under different conditions.

The usual attitude taken by experimenters when reporting their work is that random errors are sufficiently indicated by the scatter between points, and between repeat runs, as shown on the graphs of results, and that consistent errors are obvious from the description of the apparatus used: occasionally a real effort is made to estimate the accuracy of the experiment, but the most that appears in print is a bald statement of the estimated error without much indication of how it was made up. To be sure, the probable accuracy of the more common measurements is fairly well understood by most other experimenters, though possibly not by theoretical workers who read the report, but workers who are using novel techniques should try to give a full account of the errors involved.

Report Writing

Although only a very small percentage of experimental investigations is ever published, chiefly because prototype and project test results are not usually of public interest, but also because work connected with weapon development is subject to security restrictions, it is nevertheless usual for written reports of an experiment to be prepared for consumption by people who are not familiar with the experimental details, whether within the experimenter's organization or outside it.

If the work is a continuation of work previously reported it may not be necessary to give details of the apparatus, the experimental procedure or even the reasons for the enquiry, and if the report is intended only for internal consumption by people who know and trust the experimenter and his methods, no discussion of the accuracy or limitations of the results may be needed. However, even internal reports have a tendency to circulate afar, and the typical reader should be taken to be a stranger, intelligent and well-read, but suspicious by nature and not especially familiar with the subject of the experiment. Any report likely to have more than a very limited circulation should therefore begin with an introduction setting out the reasons for enquiry, briefly reviewing the current state of knowledge of the subject of the enquiry, and giving a few key references to previous work. It is not necessary to describe the whole history of the subject, though tastes differ here: a number of papers on boundary layers give, as reference 1, Prandtl's paper of 1904, which few of the authors can ever have read. It is not good practice to give as a reference a paper one has not seen: the source from which one derives the reference may have misquoted the title or page number, let alone the contents, and there are several examples in the literature of the perpetuation of such errors.

The experimental arrangements should be described in sufficient detail to be intelligible to our hypothetical reader. In particular, a clear diagram of the test arrangements should be given, with enough dimensions and cross-sectional views to define the shape accurately. Any permanent test rig like a wind tunnel should be identified, preferably by reference to a calibration report, and the size of the working section quoted, so that the reader can, if he wishes, check or modify the corrections for wind tunnel interference. The Mach number and Reynolds number—and any other relevant non-dimensional parameters—should be given, preferably with sufficient dimensional data for them to be calculated afresh. It is not usual to describe the mechanical arrangements or the design of the measuring apparatus unless there is some novelty about them, with the exception that dimensions of Pitot tubes, static tubes and other probes are commonly given because of a general uneasiness about their performance and their effect on the flow.

Some report writers prefer to point out the more interesting features of the results in a short section before the main discussion, while others combine the statement of results with the discussion of their significance. In either case, a summary should be given of the variables measured and the range of incidence, Mach number or other independent variables over which the tests were made. In many cases it is better to give results in graphs rather than tables: most people find graphs much easier to assimilate at a glance, and we must assume that our hypothetical reader is a busy man and anxious to extract the greatest amount of information from the report in the least possible time. The graphs should be prepared so that values can be read off them by interested readers: if possible, the graphs should be reproduced so that units on the scales correspond to inches, centimetres or fractions thereof.

Curve Fitting

The drawing of smooth curves through scattered readings always presents some difficulty. What is certain is that the report writer should always show the experimental points as well as the smooth curves so that his readers may exercise, if they wish, the prerogative of drawing their own curves through his points, and so that some idea is given of the scatter of the results. If the results purport to show a linear variation of one variable with respect to another, the method of least-squares can be used to fit the best straight line to the data: the procedure is to add together the squares of the perpendicular distances from each experimental point to a straight line whose equation is, say, $y = mx + c$, and differentiate the resulting sum with respect to m to find the value of m for which it is a minimum, repeating the process for c, to find the values of m and c for which the sum of the squares is least. The procedure can be extended to polynomials of best fit.

The use of these methods is tantamount to an admission that one does not know the causes of the scatter, and if it is known that the accuracy of measurement of the plotted variables is not the same over the whole range of measured values it is usually better to draw a curve by eye, giving greater weight to the points which are believed to be more accurate. Points which lie a long way from the curve outlined by the majority can be ignored in either method of determining the curve of best fit, but they should not be omitted from the graph altogether unless the reason for the discrepancy is known for certain: moderation should also be exercised in discarding complete sets of points in favour of other repeats of the same measurements, and in no case should one discard more than a small fraction of the total number of points or sets of points. Again, readers of the report should be given the opportunity to judge the improbability of the results for themselves.

The accuracy of the variables being plotted will usually vary over the range in a different way from the accuracy of the instrument readings. For instance, if one of the variables is Reynolds number based on a speed obtained from U-tube manometer readings of dynamic pressure, we expect the absolute (random) error of reading the manometer to be constant, say $\pm 0 \cdot 02$ cm of liquid, so that the percentage error of the dynamic pressures will be inversely proportional to the dynamic pressure: the percentage error of the speed and therefore of the Reynolds number will be half the percentage error of the corresponding dynamic pressure, and therefore inversely proportional to the *square* of the speed or Reynolds number.

The moral responsibility for the accuracy of the results rests with the experimenter: to be certain, in the sense of English law, "beyond all reasonable doubt", that one's results are correct to the accuracy required is a rare experience because so few experimental results can be rigorously cross-checked, and qualified confidence in the results is the most that one can normally expect. Workers with any thought for their present or future reputation will not publish results which they seriously suspect, but standards of "qualified confidence" may vary considerably.

The final section of the report should summarize the conclusions. It appears to be customary to summarize the test conditions and the conclusions at the beginning of the report as well, but this repetition scarcely seems worth while unless the summary is required for abstract cards.

A table of notation is usually necessary and should be complete down to standard symbols like ρ for density. Full definitions of any special functions used in the text should be given here, but if either a symbol or a function is only used once it should be defined in the text instead: order and method are all very well but should not be overdone. Where possible, the notation of previous workers should be used.

Style[60]

The highest standards of literary style are not required in a technical report: all that can be asked is that the style be simple and unobtrusive, and that the reader's attention shall not be distracted by grammatical lapses, infelicities, or spelling mistakes. Refined writing, of the standard that one expects in literary essays, would be out of place in a technical report, in which mellifluous synonyms for harsh-sounding words cannot be used without confusion, and in which the use of poetic words to conjure up an emotional picture would be just as distracting as a spelling mistake. Having said this, we may remark that even this minimum amount of care is frequently not taken, resulting in reports whose style presents a barrier to understanding.

The style to be cultivated is that of the news pages of the better newspapers or periodicals: there is quite a close parallel between the newspaper which has to catch the reader's eye and lead him through a news item in a logical order while enabling him to skip through it at a greater rate if he wishes, and the scientific report which should be quickly comprehensible by people to whom it is only of fringe interest, while giving a detailed and connected account of a logical investigation for the benefit of specialists in the field. Report writers, like newspaper men, must take care to separate fact and opinion: indeed the similarity only ends with the report writer's occasional need to express more complicated ideas in longer sentences than are usual in newspapers. Conscious humour or personal interjections are not in favour, and indeed any manifestation of the experimenter's feelings seems to be frowned on: irrelevant humour would transgress the rule of unobtrusiveness laid down above, but it is perhaps a pity that so few reports give any indication, intentional or otherwise, that the writer enjoyed his work, or that he was ever surprised, pleased or mystified at the results of the experiment.

Research workers, in particular, often have to invent new compound nouns or phrases to describe new phenomena. If one expects or hopes that these will pass into common usage, it is advisable to keep them short, if only because phrases with long words tend to be abbreviated to meaningless initials, a recent example being EVA for "extra-vehicular activity"; "space-walking" might have survived. Compound nouns are sometimes essential but must not be made too long: a choice specimen that actually appeared in a report (whose author would doubtless prefer to remain anonymous) is "wing upper surface intake flow momentum drag pitching moments" which means "pitching moments produced by the momentum drag of flow into an intake on the upper surface of the wing"—the natural order is exactly the opposite of the compound-noun order, and few people can mentally rearrange an unfamiliar compound of more than three or four components. Sentences should not be too long, but, if a single rather complicated idea is to be presented, a single sentence broken up by colons or other punctuation is better than a series of short sentences: the present sentence is a fair example, and the last sentence a more extreme example that might be better written as two or three sentences.

Another reason why scientific English is often turgid is that people believe that, to carry conviction, the subject and object of a sentence must both be nouns. "The observation of poorly lit dye filaments is a matter of some difficulty" is preferred to "Observing poorly lit dye filaments is difficult" or to the more natural order "poorly lit dye filaments are difficult to see". This is not quite the same as preferring "dye filaments under inadequate illumination" to "poorly lit dye filaments", which is pure pomposity.

CHAPTER 8

Specialized Branches of Fluid Mechanics

THE previous chapters treat only the more common measurement problems and techniques, with a bias towards laboratory experiments. In this chapter we discuss the problems and methods found in the different specialized branches of the subject, partly to mention some of the less common techniques and partly to show how the standard techniques are used in practice. Fluid mechanics plays only a small part in some of the disciplines mentioned below, and the present account of that small part is heavily biased towards the more unusual measurement techniques, so that it is more a collection of caricatures than a balanced review. A better overall picture of a given subject can be obtained from the specialist journals mentioned below: here there is space for only one typical experiment on each subject.

Aeronautics

1. PROBLEMS

The specialized problems of hypersonic flight and of heat transfer to re-entry vehicles cannot be usefully discussed in an elementary book such as this. The current problems in the design of manned aircraft are connected with improving the efficiency of systems whose general behaviour is fairly well understood. Examples are the development of supersonic passenger aircraft to operate at fares little greater than aircraft flying at a third of the speed, and the continuing efforts to improve the efficiency of these latter high-subsonic aircraft. Aerodynamic efficiency is almost bound to decrease as speeds increase into the transonic region, but the hope is that this will be outweighed by improvements in engine efficiency at higher intake total pressures and temperatures.

2. Experiment

Aircraft designed for shock-free flow at high subsonic Mach numbers usually have poor low-speed characteristics and complicated flap systems are used to overcome this. An illustration of the various high-lift devices used on a typical airliner, and a warning that compressibility effects may occur even at quite low nominal Mach numbers, is given by a recent paper[61] on the stalling characteristics of a Trident airliner. To get as high a Reynolds number as possible, and to permit the Reynolds number and Mach number to be varied independently, a pressurized tunnel, the R.A.E. 8 ft supersonic tunnel, was used (previous work had shown that the flow quality at low Mach numbers was still acceptable). The model was about 1·5 m in span and, except for some imperfections which were remedied during the tests, was an accurate 1/18·86 scale model of the Trident 1C. It was not thought necessary to simulate flow through the (rear-mounted) engines: this has been done in some other tests on complete aircraft by fitting a suitable ejector pump, driven by compressed air, inside the engine nacelle, but it is practically impossible to simulate all the inflow and exhaust properties exactly. In the Trident tests, the outer nacelles had straight-through ducts and the centre intake was blanked off: the whole model was mounted on a sting, attached where the central jet exhaust would be and containing a strain gauge balance. The engine flow would not be expected to affect the flow over the wing and in any case it would have been difficult to decide on a representative engine flow: normally, the engines would be at a fairly low power setting when the aircraft was in landing configuration but the pilot would naturally increase power if a stall was thought to be imminent.

The high-lift devices fitted to the aircraft are two types of leading-edge flap, double-slotted trailing edge flaps, boundary-layer fences (one on each wing), spoilers, and fifty-seven vortex generators on each wing.

Force measurements, and photographs of tufts and surface oil flow patterns, were taken over a range of Reynolds numbers up to about a third of the flight Reynolds number. Force measurements were made with and without tufts to check that the tufts did not interfere appreciably with the flow. Tufts are, of course, easier to use than oil because a range of conditions can be covered without

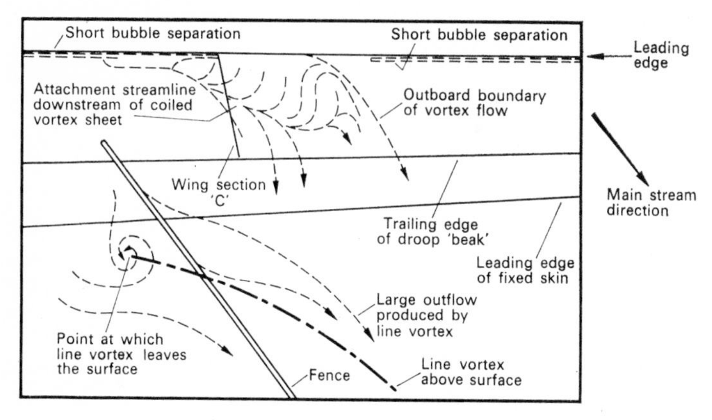

FIG. 46. Explanation of oil-film picture of Plate 3.

stopping the tunnel. Oil pictures are on the whole more informative but evidently it was felt that the increased testing time required in this large and expensive tunnel could not be justified.

The surface flow patterns obtained were very complicated and not easy to summarize. Plate 3 shows the oil flow near one of the boundary layer fences, and Fig. 46 gives a partial explanation: the centre of the spiral to the left of the fence marks the spot at which an intense vortex leaves the surface, probably spilling over the fence and causing the curious region of large outflow angle just outboard of the fence. This is, of course, only a small portion of the wing, the width of the photograph being about one-third of the chord of the wing, centred near the leading edge about one-third of the semi-span from the root.

The main conclusion of the experiment was that even the highest Reynolds number reached in this large tunnel was only just enough for the model stalling behaviour to resemble that of the full size aircraft. It was also found that the flow near the "knuckle" of the depressed leading-edge flap was quite sensitive to Mach number even at Mach numbers as low as 0·15 (local Mach numbers were not measured but must obviously have been at least 0·5), and the maximum lift coefficient changed from less than 1·7 at $M = 0{\cdot}3$ to nearly 1·9 at $M = 0{\cdot}15$: it is therefore desirable to get the aircraft Mach number right in model tests, and to achieve at least one-third of the full-scale Reynolds number if leading-edge stall is expected.

Chemical Engineering

1. Problems

The special fluid-mechanic problems of this subject are mass transfer and mixing of different fluids, especially in two-phase flow (see also p. 189). At least in one-phase flow, the physics of mass transfer is very similar to that of heat transfer, so the subject has much in common with heat transfer and combustion (see below). Turbulent transport processes are important: because of the breadth and complication of chemical engineering, most experimental studies are of particular arrangements rather than basic principles, although many measurements and techniques originating in chemical engineering are of value in other subjects. One example is the conductivity probe for concentration measurements in electrolytes. The probe consists of a very small electrode (typically the 5 μm tip of a platinum wire imbedded in glass) to which an electrical potential is applied, the "earth" electrode being some distance away: the electrical resistance of the path between the electrodes is determined almost entirely by the conductivity of the liquid very close to the small electrode, where the potential gradient is greatest. Apart from manufacture, the main difficulty is in keeping the electrode clean. Both mean and fluctuating conductivity can be measured, and the usual application is to turbulent mixing of dissimilar liquids. Other examples of the willingness of chemical engineers to exploit unusual effects for flow measurement are the electrolytic wall shear stress meter mentioned on p. 102, and the flash photolysis technique of dye production (p. 156).

Two-phase flows range from aerosols (in which hot wire anemometers have been used successfully) to gas/solid fluidized beds. Fluidized beds, in which heavy particles are suspended as a slurry in a stream of fluid moving vertically upwards, are a powerful method of increasing the surface area and mass flow available for a chemical reaction. The bed as a whole behaves as a liquid (for instance the surface returns to the horizontal after deformation). An avalanche is a naturally occurring fluidized bed.[62] The British Coal Utilization Research Association has developed a fluidized bed burner which will work with a very wide range of coals. Most of the measurements made in fluidized beds have been visualization studies to check on flow uniformity.

2. Experiment

An elegant technique for quantitative use of an apparently qualitative flow visualization method was used by Woollard and Potter[63] in an experiment on a fluidized bed. The bed consisted of glass beads and was driven by air, and the object of the experiment was to find the distribution of the solid material entrained in the wake of a single bubble injected when the main air current was adjusted so that fluidization was about to occur. For reasons not explained in the paper, previous workers had suggested that the displacement of solid material should be the same as the displacement of fluid caused by the passage of a sphere through a fluid in the case of irrotational flow.

The glass beads that formed the bottom third of the bed were coated with a mixture of dye and salt, the excess coating being removed by passing a strong current of air through the bed to fluidize it vigorously. The bed was made up with untreated beads. After the passage of the air bubble, a "spire" of dyed beads projected into the top part of the bed and some beads lay scattered on the surface. 0·5 cm thick layers of the bed were then removed one by one and dropped into water. The number of dyed beads was then inferred from the electrical conductivity of the resulting salt solution, presumably using the results of a calibration experiment with a known number (mass) of beads from the same batch.

The total volume of solids displaced above the original level was found to be about 0·3 of the bubble volume, against 0·5 for the passage of a spherical bubble in irrotational flow.

Oceanography

1. Problems

Oceanography is an easier study than meteorology, inasmuch as the ocean leaks into the atmosphere but not vice versa, but observations are even more difficult and several unique phenomena occur, in addition to the obvious one of surface waves. Buoyancy effects can be produced either by temperature differences or by differences in salt concentration (salinity): since the ocean is heated from above it is mostly thermally stable. The first few tens of metres below the surface have many features in common with the surface layer of the atmosphere: both play an important part in pollution studies. The oceanic surface layer is strongly stratified, unstable or weakly stable regions being bounded by thin, highly stable layers in which the temperature falls appreciably for a small increase in depth, the main layer that separates the surface flow from the stable depths being called the thermocline.

Current measurements rely mainly on tracer techniques, speeds being too low, and static pressure fluctuations (due to surface waves) too large, for Pitot and static tubes to be used. Windmill-type current meters can be used where the flow direction is fairly steady, as in rivers and estuaries. Reference 74 is the latest in a series of hot film measurements of oceanic turbulence.

Because of measurement difficulties, the controversy about wave generation by wind and the sources of turbulence in the surface layer is largely theoretical, but several laboratory studies have been made, with a bias towards the effect of waves on the *atmospheric* surface layer: for moderate wind speeds, the argument $u_\tau y/\nu$ in the "inner law" (Fig. 27) is replaced by gy/u_τ^2.

2. Experiment

An interesting, though perhaps untypical, oceanographic experiment (ref. 52) was done off the coast of Malta to study the structure of the thermocline. A temperature-gradient meter, consisting of two thermistors 50 cm apart on a vertical rod, was used to demonstrate the existence of the thin sheets of strong temperature gradient mentioned above: other workers had found that these sheets could sometimes be seen by divers because of the sharp change in refractive index. The shear across the layers was visualized by dropping a pellet of dye, which fell at a speed much greater than that of the current and so gave an approximation to an instantaneously injected dye line. Dye traces from a fixed source showed that the thick, weakly stable layers were turbulent. Dye released from a line source into a thin, highly stable sheet did not diffuse so quickly, and showed the presence of internal waves, confined to the neighbourhood of a sheet but of much greater wavelength than the wavelets that would result from shear instability of the sheet. If these internal waves are strong enough the shear they produce may change the existing shear enough for the sheet to become unstable: short wavelets form and then break as shown in the dye picture of Plate 6. A patch of turbulence then occurs, dispersing the sheet in its neighbourhood.

This experiment is yet another example of the interaction between quantitative measurements and flow visualization, which often proceeds in an iterative sequence: the next stage in the thermocline work is evidently a set of detailed measurements of the sheet structure, for which ref. 73 is a preparation. There may be quite close connections between the behaviour of waves in the thermocline and internal waves in the atmosphere, with application to clear-air turbulence.

Physiology

Blood is a non-Newtonian fluid, largely because of the presence of the red corpuscles, which are about 7 μm in diameter and make up about 40 per cent of the total mass: blood plasma is very nearly Newtonian (shear stress proportional to rate of strain). On the application of a shear, whole blood can apparently sustain a small yield stress before any motion takes place, but under high rates of shear the behaviour is nearly Newtonian. The behaviour in capillary veins whose diameter is only a few times that of the red cells becomes very complicated. Most of the well-known techniques of flow measurement have been used in blood, with an obvious preference for those devices that can be inserted into blood vessels by minor surgery, particularly probes that can be pushed along blood vessels deep into the body. The hot film anemometer[75] has been used successfully, and electromagnetic flowmeters[76] have been used for measurement without actually disturbing the blood vessel. Both mean and fluctuating measurements are of interest, and an immediate practical application is to the design of artificial heart components (the avoidance of transition to turbulence seems to be critical) and artificial kidney machines.

The study of particle deposition in the trachea and lungs need not be done with living animals because the lungs can be "pumped" by immersing them in water in a pressurized container. Turbulent diffusion of particles is of interest, the Reynolds number in the larger passages exceeding the critical during heavy breathing or coughing. Deposition in the smaller passages is affected by their complicated branching, and by the transfer of gases to and from the bloodstream.

A less obvious object of physiological flow studies is the personal boundary layer. Heat transfer from exposed areas of skin, whether or not augmented by perspiration, plays an important part in the body's thermal balance. Schlieren visualization has proved useful but quantitative measurement of the very low velocities involved is difficult.

I have not given an example of a physiological flow experiment because of the difficulties of medical terminology: ref. 77 is a helpful review written by an aerodynamicist, and ref. 78 an aerodynamic experiment on a physiological subject.

Civil Engineering

1. Problems

The design of chimneys, tall buildings and bridges—especially suspension bridges—is influenced by the wind loads they must withstand. The static load is not usually critical, and high factors of safety are employed by the more traditional branches of the construction industry, but care is needed to ensure that aeroelastic oscillations do not reach uncomfortable or dangerous amplitudes. The classic case is the Tacoma Narrows suspension bridge, which fluttered to destruction: ref. 64 gives some interesting background information. It is now customary to test models of proposed bridges in a wind tunnel: providing that the torsional and bending stiffness of the structure can be estimated, all that is necessary is to test a short, rigid section, spring-supported at the ends. The vortex street behind chimneys, towers and cables can excite "Aeolian tone" oscillations. *Ad hoc* tests on proposed structures use the techniques developed in aeronautical aeroelasticity, the only serious difficulty being that the Reynolds number attainable in the wind tunnel is much lower than that of the full-size structure, so that laminar separation may occur on the model but not at full scale. The results of these tests are not usually published in any easily accessible form.

Ventilation of buildings and mines presents problems of measuring mass flow at rather low speeds, and several of the techniques described in Chapter 3 were developed for these purposes. Mine safety is largely dependent on proper estimates of gas mixing, especially when one gas is lighter or heavier than air and tends to settle at the floor or roof of the passage.

Aerodynamic design of chimneys, and atmospheric pollution[65], are on the borderline between civil engineering and meteorology, and indeed any investigation of the effect of the atmosphere on structures must include consideration of the atmosphere itself. Attempts to simulate a buoyant atmosphere in a wind tunnel have been confined to fundamental studies, but devices for generating a thick (neutrally stable) boundary layer in a short tunnel working section have proved surprisingly successful and permit the effect of the mean and fluctuating flow to be studied on models of acceptably large size.

2. Experiment

A recent experiment on oscillating cylinders, described in ref. 66 which contains references to the earlier published work, was done in water. Although the phenomena are of course qualitatively the same in water as in air unless cavitation occurs, one of the relevant parameters is the ratio of a typical dynamic pressure to the modulus of elasticity of the structure, and this is likely to be larger in water than in air. Therefore, experiments may be more conveniently done in water: it is also implied that hydroelasticity is just as important as aeroelasticity, and indeed dangerous oscillations of pier columns and the like have occurred.

Circular and triangular prisms, about 4·5 cm across and about 11 diameters long, were cantilevered from a strain-gauge force and displacement balance into a water flow with a speed of about 0·85 m s^{-1}. The vortex street shedding frequencies were less than 4 Hz, and all the strain gauge outputs were recorded on magnetic tape at 3·75 in./sec and replayed at 60 in./sec so that the frequencies fell within the range of a standard frequency analyser and of an analogue multiplier originally produced for use with hot wire equipment. Measurements were made with the cylinder externally forced to oscillate at different frequencies between zero and about twice the shedding frequency. The results are given as plots against frequency of the root-mean-square fluctuating lift coefficient and of the phase angle between the lift and the displacement (on which the energy transfer from the fluid to the cylinder depends). Frequency spectra of the fluctuating lift coefficient are given for some cases: as has been found in other experiments, the spectra have peaks both at the excitation frequency and at the vortex-shedding frequency unless the two are close together, in which case the vortex frequency is changed to "lock in" to the excitation frequency. This phenomenon has been found in other non-linear systems.

Heat Transfer and Combustion

1. Problems

Heat transfer is a quantity to be maximized or minimized according to the problem considered. Because the transfer of heat by turbulence is quite closely analogous to the transfer of momentum, the methods available for reducing heat transfer are fairly obvious: one active technique is film cooling, in which cool fluid is introduced at the surface, either from discrete slots or from closely spaced perforations. Several experiments have been done in isothermal flow, using foreign gas to provide the density difference and measuring concentration by means of a thermal-conductivity cell or a gas chromatograph connected to a sampling probe: turbulent transport of heat and of contaminant are identical for small density differences, although molecular transport rates are generally different. One may sometimes be prepared to pay a heavy price in fluid-mechanic performance to reduce heat transfer, but the *augmentation* of heat transfer in heat exchangers must usually be accomplished without too much increase in pumping power: there was a controversy a few years ago about whether or not roughening the surface of a pipe would increase heat transfer for a given pumping power (as might be expected, the answer was not simple).

Measurement of overall heat transfer in pipe flow by noting the increase in fluid temperature is simple enough in principle, but local heat transfer is notoriously difficult to measure. A survey of techniques is given in ref. 67. A simple technique, used in high-speed wind tunnels, and shock tunnels, is to measure the rate of increase of surface temperature of a pre-cooled model: this requires the flow to be started, or the model inserted, very quickly, but a bonus is that the heat transfer rate can be measured for a range of wall temperatures in one experiment. Steady-state heat transfer can be deduced from the temperature difference across a wall of known thickness and thermal conductivity, but care must be taken that the thermocouples do not interfere with the flow or the heat conduction: the modern technique of depositing metal layers from metallic vapour in a vacuum can be used to make very thin surface thermocouples, and electroplating can be used if thicknesses of a few tens of μm can be accepted. Alternatively, the thermocouples can be fitted to a metal rod, normal to the surface and insulated against conduction parallel to the surface by an air gap: again, the heat transfer is inferred from the temperature difference between the ends of the rod, which usually has its own heater (at the lower end) so that the temperature at the surface end of the rod can be adjusted to the surrounding value. Nominally, these devices do not need calibration (except for the thermocouples themselves): in practice heat losses and non-uniformity of conduction may occur.

The problem of combustion is to achieve a satisfactory rate of propagation of a turbulent flame front into the surrounding fluid. Usually, the flame front is nominally stationary in space and the difficulty is to prevent it being blown away from the fuel nozzle by the oxidant stream, but in the internal combustion engine the problem of excessive flame speed or detonation may arise. Since most chemical reaction rates increase with pressure or temperature, instabilities can easily occur, the most spectacular being the several kinds of oscillatory instability found in rocket engines. Probes for regions of combustion must be cooled, usually by a recirculating water stream. Optical methods can be used for density measurement but suffer from the usual disadvantage that effects are integrated along the light path. Spectroscopy is used for temperature measurement: the apparent temperature is roughly the highest temperature anywhere on the light path. The laser anemometer seems ideal for measurements in combustion chambers but I have not yet heard of its use.

Temperatures in the cooler parts of combustion chamber walls can be inferred roughly from the (permanent) changes in colour of the metal. Heat-sensitive paints have recently appeared on the market; the paint goes through a series of irreversible colour changes as its temperature is increased, and surface temperature contours can be determined to an accuracy as good as 10°C: alternatively, heat transfer rates can be inferred from the time taken for a given colour change to be reached.

2. Experiment

An experiment on film-boiling heat transfer in pipe flow, in which the metal pipe was directly heated electrically so that the local heat transfer rate was known, is described in ref. 68. Film boiling is a condition in which boiling occurs almost continuously over the surface, in distinction to the "nucleate" boiling at local hot spots which is usual when there is no appreciable mean flow (as in a saucepan). The experiment was done with liquid nitrogen but the phenomenon is also of interest in steam generation. The nitrogen was supplied from a container pressurized with helium (it was necessary to choose a pressurizing gas with a lower boiling point than nitrogen), and a valve was used to keep the test section at about 1·7 atm (boiling temperatures depend strongly on pressure). The mass flow of gas was measured by a rotameter downstream of the test section. Copper–constantan thermocouples were welded to the outside of the test pipe (it seems to have been assumed that the temperature was uniform across the thickness of the pipe) and the pipe was then insulated so that heat transfer to the atmosphere could be neglected. The last section of the pipe was transparent so that photographs could be taken of the droplets in the outflow: previous experiments on film boiling in pipe flow had shown that droplets could persist in a superheated vapour long after sufficient heat had been supplied to vaporize all the fluid, and in the present experiment it was found that droplets could exist even though the enthalpy addition was three times the latent heat (the liquid being initially just below the boiling point). This implies that calculations assuming thermodynamic equilibrium could be seriously in error.

In order to find what fraction of the nitrogen had been evaporated by the end of the test section a small quantity of helium gas was added to the liquid nitrogen at inlet, and gas samples were withdrawn at the test section exit. The percentage helium in the gas was measured in a calibrated thermal-conductivity cell and compared to the percentage in the fully vaporized fluid at the exit of the apparatus: assuming that the liquid droplets contained negligible helium, the fraction of nitrogen evaporated follows as the ratio of the latter concentration to the former. The sample gas was withdrawn through a downstream-facing hole in a circular cylinder spanning the test section: it was presumed that only gas entered the probe but the measured concentration increased significantly with the sample flow rate so that the rate of droplet accumulation at the back of the cylinder may have been enough to cause errors at the lower flow rates.

Although the behaviour of the fluid in the upstream part of the tube may be properly termed "film boiling" with a thin layer of vapour or spray close to the walls and a homogeneous core of liquid, the latter soon entrains vapour and breaks up into a cloud of droplets. The acceleration of the fluid is considerable (the ratio of liquid velocity at inlet to gas velocity after complete vaporization being the gas/liquid density ratio) and the theory presented in ref. 68 discusses the effect of the velocity difference between gas and droplets on the heat transfer to the latter. A large drop in a gas flow will break up, and the authors assumed, on the basis of existing data, that the critical Weber number for the maximum droplet size was

$$We = \rho(\Delta U)^2 d/\sigma,$$

where ρ is the density of the gas, ΔU the relative velocity, d the droplet diameter and σ the surface tension (physically, the Weber number is the ratio of the dynamic pressure to a typical surface stress: sometimes, as on p. 12, the inverse is used).

The method of calculating percentage evaporation presented in ref. 68, allowing for droplet breakup, the variation of droplet drag coefficient with acceleration, and the direct heat transfer from the wall to the droplets, was in good agreement with the measurements. A careful investigation of the complicated phenomena occurring in film boiling was a prerequisite of a successful theory.

Hydraulics and Naval Architecture

1. Problems

The branches of hydraulics that differ most from studies of gas flows are, of course, those that deal with free-surface flows and cavitation. Mass-flow measurements are much more common than in gas flows: on small scales, the flow can be diverted into a container for a given time and then weighed; on larger scales or in nature, weirs can be used, the relation between the depth of water over the weir and the mass flow being deduced from theory, model test or calibration. The methods described in Chapter 3 have nearly all been used, tracer dilution and time-of-flight measurements being common.

Water pumps and turbines have been developed to very high efficiency, especially for electric power generation: extensive use is made of model testing. A good demonstration of the high efficiency attained, and an interesting example of engineering economics, is the "pumped storage" scheme in which water pumped up into an upper reservoir during the night runs down to generate electric power during the daytime peak load, only one electrical and one hydraulic machine being used. The total efficiency is the product of four machine efficiencies and two transmission efficiencies (water and electrical), but this apparently farcical system is more economic than building new power stations for "peak lopping".

Test rigs for hydraulic machines and propellers must operate at the right cavitation number, and most water tunnels can be run at reduced or increased pressure.[69] To reabsorb the air bubbles produced in cavitation, these tunnels usually have part of the circuit arranged at a great depth below the working section, the record being 55 m in the N.P.L. No. 2 tunnel.

Ship model testing is usually done in towing tanks, the problem of producing a wave-free surface in a high-speed flume (which has something in common with a supersonic tunnel) having been solved only recently. The choice is partly economic, because high-speed water tunnels or flumes take a lot of power, the Naval Ship Development Center (formerly David Taylor Model Basin), U.S.A. 36 in. tunnel having a 2·5 MW motor and a top speed of 26 m s^{-1} (50 kt). It is not usually necessary to measure the wave pattern round a ship except to record the waterline, so that the measurements taken and the techniques used are similar to those in wind tunnels: force balances are nearly always of the strain gauge type. Wave measurements are needed in tests of the sea-keeping ability of ships, in which self-propelled radio-controlled models are operated in wave tanks, the waves being generated by wedge-shaped plungers at one end of the tank and dissipated on "beaches" at the other end. Apart from measurements of accelerations and angles, sea-keeping tests are necessarily qualitative.

2. Experiment

Reference 70 gives a useful introduction to the literature on the physics of cavitation and includes several photographs of surfaces damaged by cavitation or impact. It has been suggested recently that cavitation damage can be caused as much by the "microjet" of liquid formed when a non-spherical gas bubble collapses as by shock waves radiating from the collapsing bubble. The author of ref. 70 suggests that the mechanism of surface damage by microjet action seems basically quite similar to damage by impinging droplets or (comparatively large) liquid jets. Erosion by impinging droplets is an important fluid-mechanic problem in its own right, because considerable damage can be caused to the leading edges of high speed aircraft flying through rain and to the low-pressure blading of steam turbines operating in wet steam (the work mentioned here was done for the French electricity authority).

The rig used to study jet impact was a disc with specimens attached to its perimeter and rotated with a tip speed of up to 65 m s^{-1} so that the specimens periodically intersected the path of water jets. The jet diameter was 5 mm: the jet speed was not stated but was implied to be small compared to the disc speed. Measurements of genuine cavitation were made in a water tunnel with speeds of up to 40 m s^{-1}. For suitable choices of flow conditions, the same curve of weight loss against exposure time was found for a soft lead–antimony alloy. In both cases, there is an "incubation period" of negligible weight loss but some surface deformation, followed by a period of low rate of damage: then the damage rate increases as deep pits appear in the surface, and finally the damage rate decreases again, evidently because the pits reach an equilibrium depth.

The effect of viscosity on jet impact damage was studied by using a mixture of water and soluble oil of unit density: viscosity could be changed by changing either the temperature or the concentration, but unfortunately the author does not compare the results on a common scale of viscosity. The general trend is for increasing viscosity to decrease the damage rate.

Meteorology

1. Problems

Much of meteorology is concerned with the behaviour of water vapour, not only in order to forecast rainfall but also because the high latent heat of water causes large rates of energy transfer: the evaporation or condensation of 1 gm of water can change the temperature of 20 m^3 of air by 1°C (alternative methods of producing such a temperature change are to accelerate the air to a speed of 50 m s^{-1} or to change its altitude by 150 m). Actual measurements of cloud properties, as distinct from simple rainfall or humidity measurements, are rather specialized, and atmospheric electricity is involved in the more violent cloud motions.

Reference 71 gives a useful introduction to the fluid mechanics of meteorology.

Clear-air turbulence—that is wind fluctuations not immediately associated with clouds or mountains—is currently of great aeronautical interest, methods for its detection being somewhat difficult to develop in the absence of conclusive explanations of its occurrence. Apart from shear turbulence near the edges of jet streams, and buoyancy-generated turbulence in unstable layers, nearly irrotational motions may occur, particularly in the form of internal gravity waves, propagating horizontally in a stably stratified flow from quite distant sources of disturbance. The most promising technique for detection seems to be the identification of the temperature fluctuations associated with the turbulence or waves: possibilities are the direct measurement of infra-red radiation or the use of the naturally illuminated horizon as a shadowgraph source.

To an increasing extent, even earthbound meteorological measurements are automatically recorded for immediate or delayed transmission to a remote station, and development of recording techniques accounts for more time and effort than methods of measurement as such. At more than a metre or two above the surface, the important velocity fluctuations are slow enough to be followed by vanes, usually a pair called a "bivane", one for direction and one for inclination: wind speed can be measured by a cup anemometer, although hot wire and sonic anemometers (using the Doppler effect to measure the component of wind velocity along a line joining a loudspeaker and microphone) have also been used. Free-flying balloons can be tracked by eye or by radar to give wind speed, and can also carry instruments with telemetry equipment. Satellites have been used for infra-red temperature measurements, but their main use at the moment is to provide world-wide pictures of cloud cover for use in forecasting. In the World Weather Watch programme, starting in the early 1970's, satellites are being used as relay stations for the interrogation of freely drifting instrumented balloons and buoys.

When aircraft are used for atmospheric turbulence measurements, it is usually assumed that the turbulence is "frozen" as the aircraft passes through it, so that a measured frequency spectrum is the true wave-number spectrum scaled by the speed of the aircraft. This assumption is more rigorously valid for a fast-moving aircraft than for a fixed instrument in the field or in the laboratory, but corrections are needed for the response of the aircraft itself to the turbulence.

Micro-meteorology, the study of the first few hundred metres of the atmosphere, uses many of the techniques developed for laboratory studies of boundary layers and turbulence, with the addition of temperature and heat flux measurements which are needed to establish buoyancy effects. One of the main objects is to see how the turbulence structure parameters, particularly the ratio of momentum diffusivity to heat diffusivity, vary with buoyancy. Unfortunately, the accuracy attainable is much poorer than in the laboratory because of the impossibility of producing steady-state conditions: the atmospheric wind comprises all frequencies of fluctuation from locally generated turbulence to daily, annual and even longer periods, and arbitrary divisions must be made according to the phenomenon being studied.

2. Experiment

A typical experiment in micro-meteorology is reported in ref. 72 and accompanying papers. Measurements were made, at heights up to 16 metres above a grass surface, of wind speed, temperature and humidity, and of the vertical transport of heat and water vapour by turbulent eddies (fluxes). To complete the heat balance, heat transfer into the ground, and radiation from the ground, were also measured.

Wind speed was measured with cup anemometers whose rotation rate was recorded on electrical impulse counters. Temperature differences between different levels were measured by pairs of platinum resistance thermometers in bridge circuits: this is a useful way of minimizing experimental errors when, as in this case, one really wants to know the gradient of a quantity. Humidity was measured by a variant of the conventional dew-point instrument: instead of finding the temperature to which a surface has to be cooled for condensation to occur, this instrument determined the temperature at which the layer of liquid condensed on a potassium chloride crystal was just in equilibrium with its surroundings (the thickness of the layer, which was a saturated solution of potassium chloride, being monitored by means of an electrical resistance meter). The advantage claimed is that the equilibrium layer of condensate is less affected by contamination (for example, by salt in the air) and more likely to approximate to the bulk properties on which calibration is based, than the fine droplets obtained in the ordinary dew-point instrument. The surface flux of water vapour was measured simply by weighing a sample of turf and assuming that other mass transfer was negligible. The flux of water vapour at a height of 4 m, which should have been very nearly the same as at ground level, was measured as the correlation between humidity fluctuations (using a wet-bulb psychrometer) and vertical velocity fluctuations: this is analogous to the measurement of turbulent shear stress (the flux of u-component momentum). The heat flux was measured similarly, as the correlation between temperature fluctuations and vertical velocity fluctuations, and conductive heat transfer into the ground was measured in terms of the temperature difference between the top and bottom of a shallow plate buried in the soil. Shear stress was not measured directly: in earlier work, the authors had inferred the surface shear stress from a wind speed measurement at 0·5 m from the surface using a value of U/U_τ (see p. 100) derived from more detailed measurements in conditions of neutral atmospheric stability and assumed to be the same independent of stability, but this assumption had been criticized by other workers. Direct measurements of the shear force on an element of the surface have been made in several meteorological experiments, and a recent device is described in ref. 73.

At first sight, the measurements made in this experiment may seem to have little connection with fluid mechanics, but turbulent transport of heat, moisture and other contaminants is at least as important in meteorology as the more familiar turbulent transport of momentum and is equally a fluid-mechanic problem.

Turbomachinery and Duct Flows

1. Problems

Although bends in pipelines and ducting rarely need to be designed carefully—most of the pressure drop occurring in the straight part of the duct—the potential problems are much the same as in turbomachine design where fluid must be persuaded to change direction without extensive regions of flow separation or large losses. The most important phenomenon, in addition to those found in nearly two-dimensional flows, is the appearance of streamwise vorticity, either the secondary flow generated by turbulent stresses (see p. 31) or the vorticity that always appears when a flow with a velocity gradient in the xy plane is forced to acquire a component of velocity in the z direction. Streamwise vorticity in a shear layer supplements the exchange of mass and momentum by viscous or turbulent stresses (usefully, in the case of "vortex generators"—see p. 179) and makes calculation of the shear layer development practically impossible except in very simple cases: fortunately, the flow not too close to the surface can be regarded as inviscid for many purposes of turbomachine design, but flow separation is controlled by the very complicated flow near the surface. Experimental studies of these highly three-dimensional shear layers are needed to derive design rules and to check the performance of a given design: detailed pressure-tube surveys are tedious, and often very difficult to arrange in the confined spaces of a turbomachine; flow visualization is also difficult to arrange but gives an immediate picture of the surface streamlines. Fundamental tests on blade design are usually made in stationary "cascades" like the cascade of corner vanes shown in Fig. 12: even here, the secondary flow in the boundary layers on the side walls at the ends of the blades may complicate the flow. In real machines, the effects of rotation and of the blade tip clearance produce further complications.

2. Experiment

An interesting example of the complicated effects produced by turning a shear flow is the flow in a mitred (sharp-cornered) pipe bend. It has recently been found[80] that the streamwise vorticity pattern in the downstream leg of the pipe is bi-stable: instead of a pair of contra-rotating vortices (expected from symmetry considerations and actually found in more gentle bends) a single vortex occurs, almost filling the pipe and changing its sense of circulation at intervals in a near-random fashion. The investigation was another example of the combined use of flow visualization and quantitative measurement. The investigators used a hot wire anemometer to check that the flow in the test rig was satisfactory, and noticed that the hot wire output was bi-stable. This is not uncommon in nominally symmetrical flows with separation: in a two-dimensional diffuser with an included angle in the region of 8 deg the flow will switch from one wall to the other, an effect that is used in fluid logic devices. However, flow visualization tests in the pipe bend showed a much more complicated flow, in which the upstream influence of the vortices in the downstream leg disturbed the separated-flow region in the corner so that the symmetrical state was unstable. The switching from one sense of circulation to the other seemed to be set off by the arrival of unusually strong turbulent eddies at the corner: because the probability of occurrence of these eddies was rather small, the frequency of switching was much less than a typical eddy frequency. Tufts and powder injection (see p. 149) were used in air, and then some dye tests were done in water, mainly to see if the instability was peculiar to the test rig (the differences between air and water were, of course, immaterial and water was chosen for the check test rig because flow visualization is generally easier).

Quantitative hot wire measurements were made, using a sensing device to discriminate between the "left-handed" and "right-handed" states. A small flag was set in the flow, with contacts arranged to close one of two electric circuits according to the sense of circulation, and a logic element was constructed to pass the hot wire signal to the frequency analyser only when the circulation was in a chosen direction, thus obtaining frequency spectra for the left-handed and right-handed states separately (there are difficulties in defining spectra for a case like this, where there would still be an input to the analyser, even if there were no true turbulence in the flow, because of the difference in "mean" velocity between the left-handed and right-handed states).

The lesson to be learned from this experiment is that the detailed behaviour of a flow cannot be inferred from measurements of gross quantities like pressure drop or mean velocity (which have been measured many times in pipe bends without the switching phenomenon being detected). In the present case, mean flow measurements alone would have indicated that the flow in the downstream leg was free of swirl.

Journals

There is no journal devoted entirely to experimental techniques in fluid mechanics: as can be seen from the list of references on pp. 211–14, papers on experimental methods are published either in the specialist research journals or in the two main journals that are wholly devoted to measurement techniques, the (American) *Review of Scientific Instruments* and the (English) *Journal of Scientific Instruments.* Very frequently, techniques developed in one branch of science can be immediately applied in a totally different branch, so that the extremely wide coverage of the "Scientific Instruments" journals does not result in negligible interest for any one reader. The same applies to the interdisciplinary research journals, especially the *Journal of Fluid Mechanics*, which publishes theoretical and experimental papers on a wide range of basic research topics: the *Journal of the American Society of Mechanical Engineers* also covers a wide range of engineering topics.

Aeronautics

A.I.A.A. Journal (U.S.A.). This is the research journal of the American Institute of Aeronautics and Astronautics: it combines theoretical and experimental work on aerodynamics and aircraft structures. Papers with special application to aircraft design or operation appear in the *Journal of Aircraft* the A.I.A.A. also publishes the *Journal of Spacecraft and Rockets* and the *Journal of Hydronautics* (see below).

Aeronautical Quarterly (Britain). This is the equivalent journal of the Royal Aeronautical Society: the R.Ae.S. also publishes the *Aeronautical Journal*, roughly the equivalent of the *Journal of Aircraft* in the standard and slant of its research papers, although it also serves as the Society's bulletin and was formerly called the *Journal of the Royal Aeronautical Society.*

Zeitschrift für Flugwissenschaften (Germany). Contains papers on research work in fluid mechanics and aircraft structures: English abstracts are provided.

Chemical Engineering

Journal of the American Institute of Chemical Engineers. According to the editors, this journal contains about 40 per cent of fluid mechanics papers and a large proportion are either wholly or partly experimental. By no means all refer exclusively to chemical engineering situations, and even the specialized papers may contain details of valuable experimental techniques.

Chemical Engineering Science also publishes a number of papers on fluid mechanics but the papers in general seem to be rather more specialized than in *J.A.I.Ch.E.*

Civil Engineering

There is no regular avenue of publication for work on building aerodynamics: most of the work is either *ad hoc* or of a sufficiently basic nature to qualify for one of the basic fluid mechanics journals. The recently established journal *Atmospheric Environment* deals with the fluid-mechanic aspects of air pollution.

Heat Transfer and Combustion

International Journal of Heat and Mass Transfer. Publishes papers on these subjects, with abstracts, in several languages. Literature surveys appear regularly. "Mass transfer" is interpreted quite liberally and a number of papers of general fluid mechanic interest are included.

Journal of Heat Transfer. This is a similar journal, published by A.S.M.E. A high proportion of the papers combine theory and experiment.

Hydraulics and Naval Architecture

Journal of the Hydraulics Division, American Society of Civil Engineers. One of several parts of the A.S.C.E. proceedings, with general papers on fluid mechanics as well as papers on flows with free surfaces and duct flows. Papers introducing new experimental techniques do not seem to appear very often.

Journal of Hydronautics. Published by the A.I.A.A., with the idea that the methods of aeronautics are also applicable to the design of high-speed boats (or, if not, should be). Papers on hydrofoil behaviour appear frequently: the journal appears to have been started because of the number of hydrofoil papers that were submitted to the A.I.A.A.'s aeronautical journals.

Meteorology

Quarterly Journal of the Royal Meteorological Society. Has a large proportion of experimental papers (the longer theoretical papers tend to go to the Journal of the Atmospheric Sciences, although the latter does publish some observational work). There are many papers dealing with subjects other than fluid mechanics but they are often interesting for the light they shed on the transport properties of the atmosphere.

Journal of Applied Meteorology. Published, like *J. Atmos. Sci.*, by the American Meteorological Society: also has a large proportion of experimental papers and papers on measurement techniques.

Oceanography

There does not appear to be a journal specializing on work in the upper ocean, which is the part of the subject most likely to interest workers in other branches of fluid mechanics. *Deep Sea Research* publishes papers on a fairly wide range of subjects, but many papers on the fluid mechanics of oceanography and air–sea interaction appear in the meteorological or interdisciplinary journals. The Russian journal *Fiziki Atmosferii i Okeana* is available in English translation as *Atmospheric and Oceanic Physics*: most of the papers describe Russian work, but some foreign papers appear (notably one reporting work sponsored by the Defence Research Board of Canada).

Turbomachinery

This subject does not appear to have its own journal: much of the work is done for aeronautical purposes and appears in aeronautical journals and report series. Papers also appear in mechanical engineering journals such as the A.S.M.E. *Journal of Engineering for Power* and the *I.Mech.E. Proceedings* and *Journal of Mechanical Engineering Science*. *The Heating and Ventilating Engineer* occasionally has papers on experimental techniques, especially flowmeters, for use in duct flows.

Abstracting Journals

These are useful for getting a quick impression of current work in a subject although the coverage of what are fringe interests for a given journal (e.g. oceanography in an aerospace journal) may be misleadingly selective.

Applied Mechanics Reviews contains *reviews* of papers on all branches of solid and fluid mechanics: the reviews are written by experts and not by the journal's staff and, probably for this reason, tend to take some time to appear.

International Aerospace Abstracts, published by the A.I.A.A., contains abstracts of published papers on a wide range of fluid mechanic and engineering subjects: there are sections devoted to research facilities and to instrumentation.

Scientific and Technical Aerospace Reports, published by N.A.S.A. in the same format as International Aerospace Abstracts, contains abstracts of unpublished reports issued by universities, firms and government establishments in all parts of the world, the only qualification being that N.A.S.A. should receive a copy. The publication delay is small (as in the case of I.A.A.) and, as more agencies outside the U.S.A. realize that it is a unique way of ensuring that their work becomes known, STAR should become an almost complete guide to research in the aerospace branches of fluid mechanics and a very valuable guide to nearly all of the common branches of fluid mechanics. Neither of these journals *reviews* papers, although the abstracts that are written by the journal staff sometimes contain comments on gross shortcomings: very frequently an attractive-sounding title and abstract conceal an irrelevant or incompetent report, and there is no cure for this short of turning the abstracting journal into a reviewing journal which, on the scale of I.A.A. and STAR, would be a very costly business. One does get to know the establishments or authors whose papers are unsatisfactory, but the student or beginner may waste some time and effort. Most of the papers appearing in STAR can be bought from the journal office in paper copy or microfiche (postcard-sized microfilm) form.

In addition to abstracting journals (of which the above is only a small selection) there are several information services run by agencies or profit-making companies. An interesting concept is the *Science Citation Index*, which lists all the papers that have been referred to in currently published journal articles: this enables one to start with an important, but dated, reference and then locate *later* papers on the same subject. At present, this service applies only to published papers and not reports.

Notation

A	Area (of tunnel cross-section, of aerofoil section); aspect ratio.
a	Speed of sound.
C	Coefficient (suffixed to indicate lift, drag, pressure, etc.); capacity.
c	Chord; specific heat.
D	Drag.
f	Frequency; a function.
g	Acceleration due to gravity.
H	δ_1/δ_2.
h	Height (of tunnel working section).
I	Current.
j	$\sqrt{(-1)}$ (used instead of i in electricity to prevent confusion with current).
k	Thermal conductivity.
L	Lift.
l	Length, rolling moment.
M	Mach number, time constant.
m	Mass flow rate, pitching moment.
p_0	Total pressure.
p	Static pressure.
R	Electrical resistance, radius of curvature.
R_e	Reynolds number.
s	Space between cascade blades.
T	Temperature.
T_0	Total temperature.
t	Time, thickness (of aerofoil section).
U, V, W	Mean velocity components in x, y, z directions.

u, v, w Fluctuating velocity components.
$\mathbf{V}$ Resultant velocity.
x, y, z Cartesian coordinates, x usually in free stream direction and y normal to the surface.

α Angle of incidence, temperature coefficient of resistance.
β $\sqrt{(1 - M^2)}$.
γ Ratio of specific heats, camber angle (angle between tangents at leading and trailing edges of aerofoil).
δ Small increment.
δ_1 Displacement thickness (eqn. 8).
δ_2 Momentum thickness (eqn. 9).
ϵ Fraction to be added to velocity measured far upstream of tunnel model to allow for blockage effect.
θ Angle.
λ Power factor (of tunnel), mean free path.
μ Viscosity, refractive index.
ν Kinematic viscosity μ/ρ.
ρ Density.
τ Shear stress, time constant.
ϕ Potential.
ω Circular frequency $2\pi f$.

Suffixes

a Atmospheric conditions.
p Constant pressure.
v Constant volume.
w Wall, surface.
0 S.T.P., or stagnation.
1 Edge of boundary layer.
∞ Conditions far from model.
$\overline{F}$ Mean of F with respect to time.
F' F per unit span of two-dimensional body.

Answers to Examples

Chapter 1.

1. The ratios of density, viscosity, specific heat and speed of sound in air and in water: the Reynolds number and Mach number based on the properties of one or other fluid: the cavitation number (see page 36): the Froude number U^2/gh where h is the depth of the pond (the speed of long surface waves is $\sqrt{(gh)}$): the Weber number $\sigma/\rho_w U^2 d$ where σ is the surface tension of the water and ρ_w its density, and d is a typical linear dimension of the stone: the ratio d/h: and γ. The cavitation number could be neglected if it were *large* (U small), and the effect of γ is likely to be small, but the Reynolds number in both air and water will be important, and the Mach number in air will determine the loudness of the splash.

2. The equations are (mass) $\gamma + \delta = 0$, (length) $\alpha + \beta - \gamma - 3\delta + \epsilon = 0$, (time) $\alpha + \gamma + \epsilon = 0$, leading to $(U/a)^{-\epsilon}(Uc\rho/\mu)^{-\gamma}$.

4. Because the streamwise gradients of velocity cannot be assumed small compared with gradients perpendicular to the surface, nor can velocities perpendicular to the surface be assumed small compared with velocities parallel to the surface, near the separation point. In the wake far enough behind the body these assumptions are again valid because the wake grows slowly with distance downstream.

5. $\lambda \propto \mu/a\rho$, leading to $\lambda \propto T^{1\cdot 26}/p$.

Chapter 2.

1. Suspend a body of known drag in the working section and find the increase of power input to the fan needed to drive the tunnel. If this is greater or less than the power needed to drive the air past the body (equal to the drag of the body times the velocity of the air divided by the approximate efficiency of the fan) it

follows that the fan is respectively too heavily or too lightly loaded, provided that the body does not interfere with the flow in other parts of the tunnel.

2. This has been done on small tunnels and has structural advantages for water tunnels where the stresses in wire screens may be high and oscillations may occur. It is difficult to buy, and expensive to make, honeycomb with the high quality of cell alignment required.

3. As the fluctuations arise from turbulence in the wall boundary layers, they could be eliminated by keeping the wall boundary layers laminar, for instance by sucking the turbulent layers away at the beginning of the working section and applying distributed suction downstream of this point so as to keep the Reynolds number based on boundary layer thickness below the value for transition to occur. There are certain practical difficulties in doing this.

4. Because round tubes are stronger than square tubes, and because the flow quality normally obtained is not high enough to warrant careful nozzle design, even if real-gas effects could be allowed for accurately.

Chapter 3.

1. 30 ft/sec, a value which may well be attained near the surface even when the tunnel speed is much higher than this.

2. $\partial p/\partial y - 2p/R = -2P/R$, or $(\partial/\partial y)(p.\exp[-2y/R])$
$= -(2/R)\,P.\exp[-2y/R]$ if $R =$ constant.

3. $-ms.\mathrm{d}T/\mathrm{d}t = (a + bU)(T - T_1) - (a + bU_1)(T_1 - T_a)$
$= (a + bU)(T - T_1) + b(U - U_1)(T_1 - T_a)$

where m and s are the mass and specific heat of the wire and a and b are (dimensional) constants. Suffix 1 represents the equilibrium conditions of temperature and velocity. If the velocity increases suddenly to U_2,

$$ms.\mathrm{d}T/\mathrm{d}t + (a + bU_2)T = (a + bU_2)T_1 - b(U_2 - U_1)(T_1 - T_a).$$

The solution to this equation is that given on p. 92 if

$$M = ms/(a + bU_2) \text{ and } \Delta T = -b(U_2 - U_1)(T_1 - T_a)/(a + bU_2).$$

4. Buoyancy force/viscous force $= -\Delta\rho g d/(\rho\nu^2/d^2) = g d^3 \Delta T/\nu^2 T$. See p. 107 for another example of the use of $\rho\nu^2/d^2$ as a typical viscous force. This number is known as the Grashof number.

5. The pressure difference could be generated between the ends of a long capillary tube in which the flow was known to be laminar and the pressure drop therefore directly proportional to the rate of flow.

Chapter 4.

1. 0·4 sec. The lag would have decreased to 1 per cent of the pressure difference after about 2 sec.

2. The ratio of meniscus movement to applied pressure measured as a liquid head is $1/(a/A + \sin\theta)$ where a and A are the normal cross-sectional areas of the small and large tubes and θ is the angle of inclination of the tube. It has been suggested that the sensitivity could be increased to very high values by making θ negative: the meniscus does not necessarily break down as soon as θ becomes negative.

3. The chief reason is the fear that the presence of the hole may upset the flow (the disturbance is, crudely speaking, equal to the pressure error times the hole area) particularly by causing transition of a laminar flow.

Chapter 5.

1. This, too, is largely a matter of custom. Null-displacement balances were used on high-speed tunnels before the advent of strain gauges, but strain gauges are preferred because of the saving of tunnel running time. Tunnel time is not such an important consideration in low-speed tunnels which absorb less power, so that null-displacement balances are adequate, with or without automatic recording facilities.

2. A sting attached to the wing of a complete aircraft model instead of the rear of the fuselage could be used: the sting could

alternatively be attached to the jet pipe. Magnetic suspensions have also been used successfully: the only difficulty apart from the provision of suitably strong magnets is the need to make the model position stable against translational or rotational disturbances in any direction.

Chapter 6.

1. The answer depends on the definition of the point from which the filament is emitted: if the point is stationary with respect to the body in each case the answer is yes.

2. The terminal velocities of such particles falling under gravity are not always small compared with the velocities in the flow (this statement combines the "gravitational" and "inertial" objections).

3. No. The evaporation rate could have increased simply because the speed of flow increases by a factor of twelve over the speed in the wide end, thus increasing the surface friction without any change necessarily occurring in the surface friction coefficient.

4. Because the rays of light are all refracted, the resulting change in the position of the rays at the screen bears no relation to the thickness of the shock wave: it can be seen that the rays passing through the region of maximum density gradient will be deflected so as to increase the illumination at larger values of y (eqn. (29)) at the expense of smaller values of y, and the separation between the light and dark areas will be of the order of $\theta_{max}\, l$ where l is the distance between the shock and the screen.

References

I HAVE included some "unpublished" references—that is reports or papers that have not appeared in journals or report series on public sale—because in recent years it has become much easier to get such material and the dividing line between "published" and "unpublished" work has become blurred. The reason for this is the spread of photocopying and micro-copying machines: one often finds that microfilm or paper copies of reports are available from government documentation agencies or libraries at a small charge whereas the organization for which the report was prepared may regard it as "unpublished". It is usual for papers of permanent interest to be published in journals even if the "unpublished" version has been quite widely circulated: publication by a reputable journal is an indication that a piece of work has been reviewed and approved by some expert in the field, and the same cannot always be said of internal memoranda or contract reports. One should always quote the published form of a reference if possible.

In nearly all cases, the author of a paper will be able to supply a copy or suggest a source of supply and references ought to be given in sufficient detail for the author or his organization to be located: as a last resort, the writer who quoted the reference should be able to say where he found it. Since the issuing organization will always have an archive copy of a report which can be photocopied, there is no reason why even unpublished papers should ever become out of print. The motive for this free interchange of scientific papers is the feeling that an individual or organization will gain far more by the scientific value of reports received than is lost by the intrinsic value of reports distributed. Since the circulation *is* an exchange, students with nothing to offer are advised to request reports through their college libraries.

Most unpublished papers can be copied: papers which are on public sale, nominally to make a profit, are protected by copyright.

1. Sutton, O. G., *Mastery of the Air*, Hodder & Stoughton, London, 1965.
2. Pankhurst, R. C. *Dimensional Analysis and Scale Factors*, Chapman & Hall, London, 1964.
3. Rosenhead, L. (Ed.), *Laminar Boundary Layers*, Clarendon Press, Oxford, 1960.
4. Benson, R. S., *Advanced Engineering Thermodynamics*, ch. 12, Pergamon, Oxford, 1967.
5. Liepmann, H. W. and Roshko, A., *Elements of Gasdynamics*, ch. 2, Wiley, New York, 1957.
6. Hinze, J. O., *Turbulence*, McGraw-Hill, New York, 1959.
7. Liepmann, H. W. and Roshko, A., *Elements of Gasdynamics*, ch. 14, Wiley, New York, 1957.
8. Pankhurst, R. C. and Holder, D. W., *Wind Tunnel Technique*, Pitman, London, 1952.
9. Pankhurst, R. C. and Gregory, N., ch. 10 of *Laminar Boundary Layers* (Ed. Rosenhead, L.), Oxford, 1963.
10. Ladenburg, R. W. *et al.*, (Ed.) *Physical Measurements in Gas Dynamics and Combustion*, vol. 9 of *High-Speed Aerodynamics and Jet Propulsion*, Princeton, 1955.
11. Donovan, A. F. *et al.* (Ed.) *Problems of High-speed Aircraft and Experimental Methods*, vol. 8 of *High-Speed Aerodynamics and Jet Propulsion*, Princeton, 1961.
12. Robertson, J. M., Water tunnels for hydraulic investigations, *Trans. A.S.M.E.* **78**, 95 (1956). A more recent but less accessible review is Gleed, D. B. and Saiva, G., A selective survey of literature on water tunnels to provide a basis for the design of the Lucas Heights water tunnel, Australian Atomic Energy Commission, Lucas Heights Research Establishment, TM 410 (1967).
13. Squire, H. B. *et al.*, The R.A.E. 4×3 ft experimental low turbulence wind tunnel. Part I: *A.R.C. R. and M.* 2690 (1948). Part II: *A.R.C.* 11,829 (unpublished). Part III: *A.R.C. R. and M.* 2905 (1951). Part IV: *A.R.C. R. and M.* 3261 (1953).
14. Bradshaw, P. and Pankhurst, R. C., The design of low-speed wind tunnels, *Progress in Aeronautical Sciences*, vol. 5, no. 1, Pergamon, London, 1964.
15. Pope, A. and Goin, K. L., *High Speed Wind Tunnel Testing*, Wiley, New York, 1965.
16. Goethert, B. H., *Transonic Wind Tunnel Testing*, AGARD*ograph* 49, Pergamon, London, 1961.
17. Wallis, R. A., *Axial Flow Fans*, Newnes, London, 1961. See also Dixon, S. L., *Fluid Mechanics and Thermodynamics of Turbomachinery*, ch. 6, Pergamon, 1966.

18. SALTER, C., Experiments on thin turning vanes, *A.R.C. R. and M.* 2469 (1946).
19. COHEN, M. J. and RITCHIE, N. J. B., Low-speed three-dimensional contraction design, *J. Roy. Aero. Soc.* **66,** 231 (1962).
20. POTTER, J. L. and CARDEN, W. H., Design of axisymmetric contoured nozzles for laminar hypersonic flow, *J. Spacecraft and Rockets* **5,** 1095 (1968).
21. LIEPMANN, H. W. and ROSHKO, A., *Elements of Gasdynamics*, Wiley, New York, 1957.
22. BATCHELOR, G. K., *Homogeneous Turbulence*, Cambridge, 1953.
23. LAUFER, J., Aerodynamic noise in supersonic wind tunnels, *J. Aero/Space Sci.* **28,** 685 (1961).
24. FERRI, A. (Ed.) *Fundamental Data Obtained from Shock Tube Experiment*, AGARD*ograph* 41, Pergamon, London, 1961.
25. ROGERS, E. W. E., BERRY, C. J. and DAVIS, B. M., Experiments with cones in low-density flows at Mach numbers near 2, *A.R.C. R. and M.* 3505 (1967).
26. BRYER, D. W. and PANKHURST, R. C., The determination of wind speed and flow direction by pressure sensing instruments, *NPL Aero. Special Report* 010 (1968).
27. OWER, E. and PANKHURST, R. C., *The Measurement of Air Flow*, Pergamon, London, 1964.
28. MACMILLAN, F. A., Experiments on Pitot tubes in shear flow, *A.R.C. R. and M.* 3028 (1956).
29. BRADSHAW, P. and GOODMAN, D. G., The effect of turbulence on static-pressure tubes, *A.R.C. R. and M.* 3527 (1968).
30. GADD, G. E., Interactions between normal shock wave and turbulent boundary layers, *A.R.C. R. and M.* 3262 (1962).
31. BRYER, D. W., WALSHE, D. E. and GARNER, H. C., Pressure probes for three-dimensional flow, *A.R.C. R. and M.* 3037 (1955).
32. British Standards Institution, Flow Measurement, B.S. 1042. See also American Society of Mechanical Engineers, *Fluid Meters Report*, 5th edn., 1959.
33. ASCOUGH, J. C., The development of a nozzle for absolute airflow measurement by pitot-static traverse, *A.R.C. R. and M.* 3384 (1963).
34. WEIR, A., YORK, J. L. and MORRISON, R. B., Two- and three-dimensional flow of air through square-edged sonic orifices, *Trans. A.S.M.E.* **78,** 481 (1956).
35. SHERCLIFF, J. A., *The Theory of Electromagnetic Flow Measurement*, University Press, Cambridge, 1962.
36. BRADSHAW, P., Thermal methods of flow measurement, *J. Sci. Instrum. Ser.* **2,** (1) 504 (1968).
37. BELLHOUSE, B. J. and SCHULTZ, D. L., The determination of fluctuating velocity in air with heated thin film gauges, *J. Fluid Mech.* **29,** 289 (1967).
38. BRADBURY, L. J. S., A pulsed wire technique for velocity measurements in highly turbulent flows, *NPL Aero Rep.* 1284 (1969).

39. HEAD, M. R. and SURREY, N. B., Low speed anemometer, *J. Sci. Instrum.* **42,** 349 (1965).
40. PATEL, V. C., Calibration of the Preston tube and limitations on its use in pressure gradients, *J. Fluid Mech.* **23,** 185 (1965).
41. MITCHELL, J. E. and HANRATTY, T. J., A study of turbulence at a wall using an electrochemical wall shear stress meter, *J. Fluid Mech.* **26,** 199 (1966).
42. SHAW, R., The measurement of static pressure, *J. Fluid Mech.* **7,** 550 (1960).
43. THOM, A. and APELT, C. J., The pressure in a two-dimensional static hole at low Reynolds numbers, *A.R.C. R. and M.* 3090 (1958).
44. CHUBB, T. W., The response of a narrow bore pressure measuring system to step and oscillatory pressures, *R.A.E. Rep.* 68010 (1968).
45. GARNER, H. C. (Ed.) Subsonic wind tunnel wall corrections, AGARD-*ograph* 109 (1966).
46. MASKELL, E. C., A theory of the blockage effects of bluff bodies and stalled wings in a closed wind tunnel, *A.R.C. R. and M.* 3400 (1963).
47. WALBRIDGE, N. L., Off-crest optical method of measuring the amplitude of surface ripples, *Rev. Sci. Instrum.* **39,** 672 (1968): author offers list of 51 references to ripple tanks.
48. CLAYTON, B. R. and MASSEY, B. S., Flow visualization in water: a review of techniques, *J. Sci. Instrum.* **44,** 2 (1967).
49. HAMA, F. R., Streaklines in a perturbed shear flow, *Phys. Fluids* **5,** 644 (1962).
50. HIGNETT, E. T., The use of dust deposition as a means of flow visualization, *A.R.C. C.P.* 651 (1962).
51. MALTBY, R. L. (Ed.) Flow visualization in wind tunnels using indicators, AGARD*ograph* 70 (1962).
52. WOODS, J. D., Wave-induced shear instability in the summer thermocline, *J. Fluid Mech.* **32,** 791 (1968).
53. BLAND, R. E. and PELICK, T. J., The schlieren method applied to flow visualization in a water tunnel, *Trans. A.S.M.E.* **84D,** 587 (1962).
54. PIERCE, D., Photographic evidence of the formation and growth of vorticity behind plates accelereted from rest in still air, *J. Fluid Mech.* **11,** 460 (1961).
55. MOWBRAY, D. E., The use of schlieren and shadowgraph techniques in the study of flow patterns in density stratified liquids, *J. Fluid Mech.* **27,** 595 (1967).
56. SCHRAUB, F. A., KLINE, S. J., HENRY, J., RUNSTADLER, P. W. and LITTELL, A., Use of hydrogen bubbles for quantitative determination of time-dependent velocity fields in low speed water flows, *Trans. A.S.M.E.* **87D,** 429 (1965).
57. POPOVICH, A. T. and HUMMEL, R. L., A new method for non-disturbing turbulent flow measurements very close to a wall, *Chem. Engng Sci.* **22,** 21 (1967).
58. TANNER, L. H., The design and use of interferometers in aerodynamics, *A.R.C. R. and M.* 3131 (1959).

59. TANNER, L. H., Some laser interferometers for use in fluid mechanics, *J. Sci. Instrum.* **42,** 834 (1965).
60. GOWERS, E., *The Complete Plain Words*, Penguin, London, 1968.
61. ISAACS, D., Wind tunnel measurements of the low speed stalling characteristics of a model of the Hawker-Siddeley Trident 1C, *A.R.C. R.* and *M.* 3608 (1969).
62. PRANDTL, L., *Essentials of Fluid Dynamics*, Blackie, London, 1953 (p. 368).
63. WOOLLARD, I. N. M. and POTTER, O. E., Solids mixing in fluidized beds, *J.A.I. Ch. E.* **14,** 388 (1968).
64. VON KÁRMÁN, TH. and EDSON, L., *The Wind and Beyond*, Little, Brown, Boston, 1967.
65. SCORER, R. S., *Air Pollution*, Pergamon, Oxford 1968.
66. PROTOS, A., GOLDSCHMIDT, V. W. and TOEBES, G. H., Hydroelastic forces on bluff cylinders, *J. Basic Engng.* **90D,** 378 (1968).
67. BAYLEY, F. J. and TURNER, A. B., Bibliography of heat-transfer instrumentation, *A.R.C. R. and M.* 3512 (1966).
68. FORSLUND, R. P. and ROHSENOW, W. M., Dispersed flow film boiling, *J. Heat Trans.* **90C,** 399 (1968). See also MCGINNIS, F. K., III, and HOLMAN, J. P., Individual droplet heat-transfer rates for splattering on hot surfaces, *Int. J. Heat Mass Trans.* **12,** 95 (1969).
69. HOLL, J. W. and WOOD, G. M. (Eds.) *Symposium on Cavitation Research Facilities and Techniques*, A.S.M.E., New York, 1964.
70. CANAVELIS, R., Jet impact and cavitation damage, *J. Basic Engng.* **90D,** 355 (1968).
71. SCORER, R. S., *Natural Aerodynamics*, Pergamon, London, 1965.
72. SWINBANK, W. C. and DYER, A. J., An experimental study in micrometeorology, *Quart. J. Roy. Met. Soc.* **93,** 494 (1967).
73. BRADLEY, E. F., A shearing stress meter for micrometeorological studies, *Quart. J. Roy. Met. Soc.* **94,** 380 (1968).
74. GRANT, H. L., MOILLIET, A. and VOGEL, W. M., Some observations of the occurrence of turbulence in and above the thermocline, *J. Fluid Mech.* **34,** 443 (1968).
75. BELLHOUSE, B. J. and BELLHOUSE, F. H., Thin-film gauges for the measurement of velocity or skin friction in air, water or blood, *J. Sci. Instrum.*, Ser. 2, **1,** 1211 (1968).
76. WYATT, D. G., The electromagnetic blood flowmeter, *J. Sci. Instrum.*, Ser. 2, **1,** 1146 (1968).
77. ROGERS, E. W. E., Medical Fluid Dynamics: notes and thoughts on some current research activities, *NPL Aero Rep.* 1061 (1967).
78. BELLHOUSE, B. J. and TALBOT, L., The fluid mechanics of the aortic valve, *J. Fluid Mech.* **35,** 721 (1969).
79. HORLOCK, J. H., Some recent research in turbo-machinery, *Proc. Inst. Mech. Engrs.* **182,** 571 (1967–8).
80. TUNSTALL, M. J. and HARVEY, J. K., On the effect of a sharp bend in a fully developed turbulent pipe-flow, *J. Fluid Mech.* **34,** 594 (1968).

Index

(Page numbers in **bold** type indicate the main reference to each title)